ENERGY SCIENCE, ENGINEERING AND TECHNOLOGY

ELECTRIC VEHICLES AND THE WIND INDUSTRY

EFFECTS OF FEDERAL TAX CREDITS

ENERGY SCIENCE, ENGINEERING AND TECHNOLOGY

Additional books in this series can be found on Nova's website under the Series tab.

Additional e-books in this series can be found on Nova's website under the e-book tab.

ENERGY POLICIES, POLITICS AND PRICES

Additional books in this series can be found on Nova's website under the Series tab.

Additional e-books in this series can be found on Nova's website under the e-book tab.

ENERGY SCIENCE, ENGINEERING AND TECHNOLOGY

ELECTRIC VEHICLES AND THE WIND INDUSTRY

EFFECTS OF FEDERAL TAX CREDITS

ADAM P. REYES
EDITOR

NOVINKA

New York

For permission to use material from this book please contact us:
Telephone 631-231-7269; Fax 631-231-8175
Web Site: http://www.novapublishers.com

NOTICE TO THE READER

Additional color graphics may be available in the e-book version of this book.

LIBRARY OF CONGRESS CATALOGING-IN-PUBLICATION DATA

ISBN: 978-1-62417-298-4

Published by Nova Science Publishers, Inc. † New York

Contents

PREFACE

This book explores the effects of federal tax credits on the electric vehicle and wind industries. The federal government has adopted several policies to encourage the production and purchase of electric vehicles, which run partly or entirely on electric power stored in a battery that can be recharged from a standard home outlet. Federal policies to promote their manufacture and purchase include tax credits for buyers of new electric vehicles, financial support for the industry that produces them, and programs that promote efforts to educate consumers about electric vehicles and improve the infrastructure for charging them. Additionally, U.S. wind projects that use large turbines, greater than 100 kilowatts, are eligible to receive federal tax incentives in the form of production tax credits (PTC) and accelerated depreciation. Originally established in 1992, the PTC has played a role in the evolution and growth of the U.S. wind industry.

Chapter 1 - The federal government has adopted several policies to encourage the production and purchase of electric vehicles, which run partly or entirely on electric power stored in a battery that can be recharged from a standard home outlet. Such vehicles are fairly new, having been reintroduced commercially in the United States late in 2010.

Federal policies to promote their manufacture and purchase include tax credits for buyers of new electric vehicles, financial support for the industry that produces them, and programs that promote efforts to educate consumers about electric vehicles and improve the infrastructure for recharging them. The Congressional Budget Office (CBO) estimates that such policies, some of which also support other types of fuel-efficient vehicles, will have a total budgetary cost of about $7.5 billion through 2019.

Of those federal incentives, the tax credits for buying electric vehicles—which account for about one-fourth of that budgetary cost—are likely to have the greatest impact on vehicle sales. The tax credits apply to the first 200,000 electric vehicles sold by each manufacturer for use in the United States, after which the credits gradually phase out.

The electric vehicles that are the focus of this study fall into two broad classes: Plug-in hybrid vehicles are powered by an internal combustion engine, which runs on gasoline or other liquid fuels, and by an electric motor, which is powered in part by an externally rechargeable battery. (Traditional hybrid vehicles, which have been available for about 10 years, also have internal combustion engines and electric motors, but their batteries cannot be recharged externally.); All-electric vehicles, also known as battery electric vehicles, run entirely on battery power.

Chapter 2 - U.S. wind projects that use large turbines—greater than 100 kilowatts (kW)—are eligible to receive federal tax incentives in the form of production tax credits (PTC) and accelerated depreciation. Originally established in 1992, the PTC has played a role in the evolution and growth of the U.S. wind industry. Under existing law, wind projects placed in service on or after January 1, 2013, will not be eligible to receive the PTC incentive. Industry proponents are advocating for an extension of PTC availability, citing employment, economic development, and other considerations as justification for the extension. While a PTC extension may improve the prospects for U.S. wind development and manufacturing next year and beyond, the wind industry is influenced by a number of other factors. It is uncertain how the near- or long-term availability of the PTC incentive—in isolation of changes to other market factors—would either grow or sustain current wind development and manufacturing levels.

For 2012, the pending expiration of the wind PTC is actually creating a short-term surge in wind project development and related investment and employment. Wind installations in 2012 are expected to range somewhere between 10 to 12 gigawatts (GW)—a record year for the industry. However, market estimates for new installations in 2013 range from 1-2 GW if the PTC expires and 2-4 GW if the PTC is extended. Limited market activity in 2013 is partially explained by the uncertain nature of the PTC, which results in reduced manufacturing orders and development activity as developers and investors wait for official policy direction. Wind installation projections for 2014 and beyond vary with the assumed availability, and duration, of PTC incentives. However, all projections reviewed for this report expect annual

U.S. wind turbine demand to be less than the existing U.S. turbine manufacturing capacity—approximately 13 GW per year.

Other factors that can affect wind development include (1) state renewable portfolio standards (RPS), (2) U.S. electricity demand growth, and (3) the price of natural gas. State RPS policies have been the primary demand creator for wind projects, in most cases, by requiring certain utilities to source a percentage of their retail electricity sales from renewable generators. Market analysis indicates that incremental RPS-driven demand for all sources of renewable power is estimated to be 4 GW-5 GW annually until 2025. Additionally, U.S. electricity demand growth is expected to be modest for the foreseeable future, meaning that there will likely be modest demand for new electric power capacity. Finally, the price of natural gas can also influence wind markets. Low natural gas prices can erode the economic competitiveness of wind electricity, while high natural gas prices can result in opportunities for wind to compete economically without the PTC. Current estimates from the U.S. Energy Information Administration (EIA) project sustained low, but increasing, natural gas prices for the next several years.

By the end of 2012, Congress will either allow the PTC incentive to expire or it may choose to extend or modify the incentive. Should Congress decide to extend the availability of wind PTC incentives, the duration (e.g., two years, four years, permanent) of such an extension will likely be part of the policy debate. Generally, the shorter the extension the greater the short-term economic and employment activity as developers and investors accelerate development plans in order to qualify for the PTC incentive. However, this development acceleration is likely to reduce future RPS-driven demand. A permanent PTC is also a policy option that may be considered, and EIA estimates indicate that such a policy may actually reduce near-term wind capacity additions, with annual installations peaking at 4 GW in the 2030 timeframe. Higher natural gas prices, more aggressive RPS policies, and increased U.S. electricity demand could change this outlook.

In: Electric Vehicles and the Wind Industry ISBN: 978-1-62417-298-4
Editor: Adam P. Reyes

Chapter 1

EFFECTS OF FEDERAL TAX CREDITS FOR THE PURCHASE OF ELECTRIC VEHICLES*

Congressional Budget Office

SUMMARY

The federal government has adopted several policies to encourage the production and purchase of electric vehicles, which run partly or entirely on electric power stored in a battery that can be recharged from a standard home outlet. Such vehicles are fairly new, having been reintroduced commercially in the United States late in 2010.

Federal policies to promote their manufacture and purchase include tax credits for buyers of new electric vehicles, financial support for the industry that produces them, and programs that promote efforts to educate consumers about electric vehicles and improve the infrastructure for recharging them. The Congressional Budget Office (CBO) estimates that such policies, some of which also support other types of fuel-efficient vehicles, will have a total budgetary cost of about $7.5 billion through 2019.

Of those federal incentives, the tax credits for buying electric vehicles—which account for about one-fourth of that budgetary cost—are likely to have the greatest impact on vehicle sales. The tax credits apply

---

* This is an edited, reformatted and augmented version of Congress of the United States, Congressional Budget Office, dated September 2012.

to the first 200,000 electric vehicles sold by each manufacturer for use in the United States, after which the credits gradually phase out.

The electric vehicles that are the focus of this study fall into two broad classes:

- Plug-in hybrid vehicles are powered by an internal combustion engine, which runs on gasoline or other liquid fuels, and by an electric motor, which is powered in part by an externally rechargeable battery. (Traditional hybrid vehicles, which have been available for about 10 years, also have internal combustion engines and electric motors, but their batteries cannot be recharged externally.)
- All-electric vehicles, also known as battery electric vehicles, run entirely on battery power.

Tax Credits and the Cost - Competitiveness of Electric Vehicles

At current vehicle and energy prices, the lifetime costs of an electric vehicle are generally higher than those of a conventional vehicle or traditional hybrid vehicle of similar size and performance, even with the tax credits. That conclusion takes into account both the higher purchase price of an electric vehicle and the lower fuel costs over the vehicle's life.

For example, an average plug-in hybrid vehicle (that is, an electric version of the typical light-duty vehicle) with a battery capacity of 16 kilowatt-hours (kWh) would be eligible for the maximum tax credit of $7,500.[1] However, that vehicle would require a tax credit of more than $12,000 to have roughly the same lifetime costs as a comparable conventional or traditional hybrid vehicle.

The additional tax credit that would be required for cost-competitiveness is smaller for electric vehicles that have small batteries or that are substituting for vehicles with low fuel economy.

Assuming that everything else is equal, the larger an electric vehicle's battery capacity, the greater its cost disadvantage relative to conventional vehicles— and thus the larger the tax credit needed to make it cost-competitive. All-electric vehicles are closer than plug-in hybrids to being cost-competitive with conventional vehicles, for a given battery size.

Tax Credits and the Cost to the Government of Reducing Gasoline Use and Emissions

The tax credits for electric vehicles have multiple direct and indirect effects on the total amounts of gasoline consumed and greenhouse gases emitted by the U.S. transportation sector. The direct effect of the credits is to subsidize purchases of electric vehicles—some of which are additional purchases and some of which are purchases that would have been made even without the credits. In itself, that direct effect leads to lower gasoline consumption and fewer emissions than would otherwise be the case. The cost to the federal government of those reductions can vary widely. For example, by CBO's estimate, the cost of the credits' direct effect on gasoline consumption ranges from about $3 to $7 per gallon saved when people buy an electric vehicle that is similar in size and performance to a conventional vehicle with average fuel economy, depending on the electric vehicle's type and battery size. The cost per metric ton of carbon dioxide equivalent (CO_2e) emissions reduced can vary even more widely—from $230 to $4,400 in CBO's estimates for electric vehicles that are comparable to average-fuel-economy conventional vehicles—because that cost also depends on the emissions released in generating the electricity used to recharge vehicles' batteries.[2] The cost per gallon of gasoline saved or per metric ton of emissions reduced is higher when electric vehicles substitute for high-fuel-economy vehicles, which use comparatively little gasoline themselves, and lower when the alternative to electric vehicles is low-fuel-economy conventional vehicles.

Because of their other, indirect effects, however, the tax credits will have little or no impact on the total gasoline use and greenhouse gas emissions of the nation's vehicle fleet over the next several years. As a result, the cost per gallon or per metric ton of any such reductions will be much greater than the amounts described above. In particular, as automakers seek to comply with the rising federal standards that govern the average fuel economy of their vehicle fleets, they can use increased sales of highfuel-economy electric vehicles as an opportunity to boost their sales of low-fuel-economy vehicles as well. (The tax credits may also lead to more sales of traditional hybrids and high-fuel-economy conventional vehicles by prompting sellers to reduce their prices in an attempt to compete with electric vehicles. But that price-competition effect also will probably be offset by greater sales of low-fuel-economy vehicles.) Consequently, given corporate average fuel economy (CAFE) standards that are high enough to constrain automakers' production decisions, the tax credits cannot significantly affect total gasoline use or greenhouse gas emissions by vehicles during the period when those standards are in effect. Currently, that period extends through 2021 for the fuel economy standards and through 2025 for the emissions standards. Previously, CAFE standards were in place for new vehicles through model year 2016, but regulators recently set more-stringent fuel economy standards that are due to take effect in 2017 and continue rising through 2021, reaching about 40 miles per gallon in that year. Parallel restrictions have been established on the greenhouse gas emissions of vehicles for model years 2017 to 2025.[3]

Over the longer term, the tax credits can affect gasoline consumption and emissions if future revisions to the CAFE standards are influenced by current sales of electric vehicles and expectations about future sales. Moreover, if the credits play an important role in helping the U.S. electric vehicle industry become self-sustaining, their effect on vehicle sales might continue to affect CAFE standards—and the resulting amounts of gasoline use and emissions—for many years after the tax credits themselves have run out.

Comparing the Tax Credits with Other Recent Subsidy Programs in the Transportation Sector

CBO compared the effects of the current tax credits in reducing gasoline use and greenhouse gas emissions with the effects of three other recent subsidy programs aimed at the transportation sector: federal tax credits for the purchase of traditional hybrid vehicles, which were in effect until 2011; federal tax credits, most of which have expired, for companies that blended biofuels with petroleum fuels; and the 2009 "Cash for Clunkers" program, which made payments to people who traded in eligible lower-fuel-economy vehicles for higher-fuel-economy vehicles.[4] Like the current credits for electric vehicles, the credits for traditional hybrids did not reduce gasoline use or greenhouse gas emissions in the short term, because sales of those high-fuel-economy vehicles allowed vehicle manufacturers to sell more low-fuel-economy vehicles and still comply with CAFE standards. By contrast, the other two programs did reduce total gasoline use and greenhouse gas emissions in the short term. The biofuel credits lowered the emissions of vehicles already purchased, and "Cash for Clunkers" raised the average fuel efficiency of all vehicles in operation (by reducing the number of less fuel-efficient older vehicles in favor of those with higher fuel economy).

The overall cost-effectiveness of such programs also depends on their long-term impact on gasoline use and greenhouse gas emissions. "Cash for Clunkers" probably did not have an ongoing influence on buyers' vehicle choices, so it did not have any additional effects over the long term. The biofuel credits may have had long-term effects by making the biofuel industry more viable in the future. And the tax credits for electric vehicles and traditional hybrids could have a long-lasting influence on CAFE standards, depending in part on the extent to which they have affected or will affect regulators' expectations about the future viability of their respective industries.

The credits' influence on future CAFE standards could also depend on the extent to which they spur additional sales of high-fuel-economy vehicles not eligible for the credits by prompting manufacturers and retailers to reduce the prices of those competing vehicles.

Possible Approaches for Future Policies

Lawmakers concerned about gasoline consumption or greenhouse gas emissions could take a number of approaches in setting future policies, some within the framework of the existing tax credits and others involving very different policies. Some lawmakers have proposed altering the dollar amounts of the tax credits or the number of vehicle purchases to which they would apply.[5] Increasing the size of the tax credits would raise federal budgetary costs, whereas reducing or eliminating the credits would reduce costs. Such changes would have little, if any, effect on gasoline use or greenhouse gas emissions over the short term, because automakers would still have to meet existing CAFE standards. However, changes in the size of the tax credits could affect future CAFE standards by influencing regulators' expectations about future sales of electric vehicles and could also affect the commercial viability of the U.S. electric vehicle industry. Increases or moderate reductions in the number of credits available would probably have little near-term impact on the credits' benefits to society—or on the cost to the government of those benefits—because sales of electric vehicles are years away from reaching the 200,000-per- manufacturer milestone now in effect. The long-term costs and benefits of changing the number of credits available are uncertain; they depend greatly on how future CAFE standards are set and on the pace at which the electric vehicle industry develops. Another option available to policymakers is to increase the federal excise tax on sales of gasoline. Raising that tax (and thus the price of fuel) would tend to reduce gasoline use and emissions, particularly after drivers had enough time to adjust their commuting patterns or to purchase different vehicles. Other policies—such as an economy-wide cap-and-trade program for greenhouse gas emissions or a tax on the carbon content of fossil fuels—could focus on low-cost reductions in emissions outside the transportation sector. Such policies would tend to minimize the total cost of achieving a given reduction in emissions, but they would probably have less of an effect on gasoline consumption than electric vehicle tax credits that achieved the same reduction in emissions, and they would have little or no effect on the development of the electric vehicle industry.

ELECTRIC VEHICLES AND GOVERNMENT POLICIES THAT SUPPORT THEM

In recent years, automakers have put more emphasis on developing vehicles that run on electricity stored in a rechargeable battery, anticipating that high gasoline prices and concern about greenhouse gas emissions will spur greater demand for electric vehicles. A limited number of such vehicles have been available in the United States in the past (General Motors offered the EV1 for lease in the 1990s, for example), but automakers began introducing a new generation of electric vehicles late in 2010. Since then, roughly 40,000 of those vehicles have been sold, out of total U.S. sales of more than 15 million light-duty vehicles. Most analysts expect sales of electric vehicles to grow modestly in coming years as automakers introduce a wider variety of models, although such vehicles will probably continue to make up only a small share of total vehicle sales for many years to come. Unlike conventional vehicles and traditional hybrid vehicles—which run on gasoline, gasoline blends that contain up to 85 percent ethanol (E85), or diesel fuel—electric vehicles use electricity stored in the vehicle's battery for some or all of the distance they travel. Once the battery is depleted, it can be recharged by being plugged into a standard home outlet. Electric vehicles fall into two broad classes:

- *Plug-in hybrid vehicles* can operate on gasoline and on stored electricity, which allows them to be driven for as many miles as conventional vehicles. Plug-in hybrids differ from the more common traditional hybrids in that they use an externally rechargeable battery to power their electric motor, in addition to an internal combustion engine. (For more details, see Box 1.)
- *All-electric vehicles* run entirely on battery power, so they cannot be driven once the battery is depleted. For that reason, all-electric vehicles generally have a larger battery than plug-in hybrids do. However, all-electric vehicles are likely to be used mainly for limited-distance travel. Long-distance travel requires frequent recharging, which is difficult because recharging can take hours.

The federal government promotes the production and purchase of electric vehicles through a number of policies. Incentives for production include grants to companies that manufacture batteries and other components of electric vehicles, subsidized loans to establish or expand facilities that produce various types of high-fuel-economy vehicles, and investment tax credits for developing facilities that manufacture clean energy technologies. Federal regulations—especially recent increases in corporate average fuel economy (CAFE) standards for cars and light trucks—also encourage the production of electric vehicles by requiring higher average fuel economy for new vehicles.[1] Among the incentives for purchasing electric vehicles, the ones with the largest impact on vehicles' price are the federal tax credits of $2,500 to $7,500 for people who buy new electric vehicles.

In addition to those federal policies, many states offer incentives for electric vehicles, such as tax credits, exemptions from state and local taxes, and preferential access to highoccupancy-vehicle lanes.[2]

Box 1. How Electric Vehicles Differ from Conventional and Traditional Hybrid Vehicles

The fundamental difference among conventional vehicles, traditional hybrids, and electric vehicles is whether they are propelled by an internal combustion engine, an electric motor, or a combination of the two. Conventional vehicles are at one end of the spectrum.

They use a battery for starting but receive all of their propulsion from an internal combustion engine—the dominant vehicle technology for the past 100 years. An internal combustion engine burns liquid fuels (primarily petroleum fuels mixed with biofuels or other blending components) and uses the energy released to power the vehicle. At the other end of the spectrum are all-electric vehicles (such as the Nissan Leaf).

They are propelled only by an electric motor using electricity from power plants or other sources of generation (such as solar or other renewable power) that is stored in a large rechargeable battery.[1] (The Leaf's battery has a capacity of 24 kilowatt-hours.) Once such a battery is depleted, the driver can recharge it using a standard plug and outlet at home, work, or elsewhere.

The battery in an all-electric vehicle is also recharged somewhat while the vehicle is operating, through a process known as regenerative braking, which captures some of the energy released when the brakes are applied. Regenerative braking is most effective during city driving, which typically involves numerous stops and starts.

Unlike conventional and all-electric vehicles, which use either an internal combustion engine or an electric motor, traditional hybrids use both. Traditional gas-electric hybrids (such as some models of the Toyota Prius) rely on an electric motor when operating at low speeds and sometimes during acceleration (in conjunction with the internal combustion engine).The electric motor boosts the vehicle's efficiency during those times and also allows the use of a smaller, more fuel-efficient engine that does not need to shoulder all of the demands of vehicle acceleration. However, traditional hybrid vehicles cannot be recharged with a plug; instead, the internal combustion engine and energy recovered from regenerative braking maintain the battery's charge and provide power to the electric motor. Thus, like conventional vehicles, traditional hybrids depend entirely on liquid fuels for their propulsion. Plug-in hybrid vehicles (such as the Chevrolet Volt and the Toyota Prius Plug-in Hybrid) have elements of electric vehicles, conventional vehicles, and traditional hybrids. Like all-electric vehicles, they have large batteries that can be recharged using a standard home outlet. (The Volt's battery has a capacity of 16 kilowatt-hours.) But like conventional and traditional hybrid vehicles, plug-in hybrids also have internal combustion engines. Plug-in hybrids run on electric power for as long as their batteries have a sufficient charge. Once the charge falls to about 30 percent of total capacity, the vehicle's internal combustion engine takes over to provide power and prevent the charge from declining further. Thus, when plug-in hybrids run on battery power, they operate like all-electric vehicles; when they use their internal combustion engine and run on liquid fuels, they operate like traditional hybrids. In some cases, plug-in hybrid vehicles can run on liquid fuels and electric power simultaneously (depending on the model and the type of driving being done), which lengthens the distance they can travel before exhausting their battery power.

1. "All-electric vehicles" can refer to other types of vehicles that run only on an electric motor, such as fuel cell vehicles, which are powered by electricity produced in the vehicle instead of electricity produced externally and stored in a rechargeable battery. In this study, however, "all-electric vehicles" refers only to vehicles that run exclusively on electricity and have a battery that is recharged from the electricity grid using a plug.

This study by the Congressional Budget Office (CBO) focuses on the federal tax credits for electric vehicles. It assesses the extent to which the credits promote sales of such vehicles by making them more cost-competitive with other vehicles. The study also examines how cost-effective the tax credits are at meeting energy and environmental goals by reducing gasoline consumption and greenhouse gas emissions. In addition, the analysis looks at how the tax credits and other federal incentives might affect the electric vehicle industry over the long term, at the possible impact of modifying the credits, and at how the costs and benefits of the credits compare with those of other policy approaches.

Federal Tax Credits Supporting the Sale of Electric Vehicles

The American Recovery and Reinvestment Act of 2009 (ARRA) created a federal income tax credit for people who purchase new electric vehicles. The size of the credit depends on the capacity of the vehicle's battery, measured in terms of kilowatt-hours (kWh) of electric power. The credit begins at $2,500 for an electric vehicle with a 4 kWh battery and increases by $417 for every additional kWh of capacity, up to a maximum of $7,500 (for a vehicle with a 16 kWh or larger battery). Earlier tax credits of as much as $3,400 were available to buyers of traditional hybrid vehicles; they expired at the end of 2010.

The tax credit is subtracted from the amount of federal income tax that the buyer owes. The credit is not refundable, however; if the amount of the credit exceeds that tax liability, the buyer does not receive the difference as a tax refund. Thus, people with relatively little income tax liability may be eligible to receive only a fraction of the credit's nominal value.

The current tax credits apply in full to the first 200,000 electric vehicles sold by each manufacturer for use in the United States, after which they will gradually be phased out. For each manufacturer, the phaseout period will begin two calendar quarters after the 200,000-vehicle threshold is achieved and last for four quarters, during which buyers of new vehicles will receive progressively smaller credits. Afterward, the tax credits will not be available for any electric vehicle made by that manufacturer, although they may still be available for electric vehicles produced by other automakers. Under current law, there is no expiration date for the credits and no limit on the number of eligible vehicle manufacturers.

Currently, three models (the Chevrolet Volt, Nissan Leaf, and Toyota Prius Plug-in Hybrid) account for nearly all of the sales of electric vehicles in the United States. Some additional models are also available, and others are expected to be introduced in the next few years. Because only about 40,000 electric vehicles have been sold for use in the United States since late 2010 (when the Volt was introduced), no manufacturer is near the threshold at which the tax credits will begin to be phased out.

The tax credits for electric vehicles could have a budgetary cost of as much as $1.5 billion (200,000 vehicles times a maximum of $7,500 per vehicle) or more for each manufacturer over a number of years, depending on the number of vehicles sold during the phaseout period. When the tax credits were being considered by the Congress, the staff of the Joint Committee on Taxation estimated that their budgetary cost would total $2.0 billion between fiscal years 2009 and 2019.[3]

Other Federal Incentives for the Sale or Production of Electric Vehicles and Related Technologies

Various other federal incentives—in the form of loans or grants—apply to the electric vehicle industry (see Table 1). For example, the Electric Drive Vehicle Battery and Component Manufacturing Initiative provides grants to support the expansion of manufacturing plants in the United States that specialize in producing batteries and other parts for electric vehicles. ARRA provided $2 billion in funding to the Department of Energy (DOE) for grants under that program. Of that amount, $1.5 billion was awarded to battery producers, intermediate suppliers for those producers, and recyclers of vehicle batteries; the other $500 million was awarded to manufacturers of components for electric vehicles and intermediate suppliers of that manufacturing.[4] DOE estimates that the $2 billion in grants will support a productive capacity of about 500,000 vehicle batteries per year by 2015.[5]

DOE's Transportation Electrification Initiative has made commitments for $400 million in grants for demonstration, deployment, and education projects involving electric vehicles.[6] Also funded by ARRA, the initiative is intended to enhance the appeal of electric vehicles to consumers by promoting awareness of the vehicles and expanding the infrastructure for charging them. Although such projects may spur sales of electric vehicles and help increase the availability of charging opportunities (thus alleviating a concern of some would-be purchasers), they do little to reduce one of the biggest impediments to electric vehicle sales—the high price of such vehicles relative to the price of nonelectric alternatives.

Table 1. Federal Incentives Available to Buyers or Producers of Electric Vehicles

Incentive	Description	Budgetary Cost (Billions of dollars)
Tax Credits for New Plug-in Electric Drive Motor Vehicles	Tax credits of up to $7,500 for buyers of new electric vehicles	2.0[a]
Electric Drive Vehicle Battery and Component Manufacturing Initiative	Grants to manufacturers of batteries and other parts for electric vehicles	2.0[b]
Transportation Electrification Initiative	Grants to establish development, demonstration, evaluation, and education projects to accelerate the introduction and use of electric vehicles	0.4[b]
Advanced Technology Vehicles Manufacturing Program	Up to $25 billion in direct loans to manufacturers of automobiles and automobile parts to promote the production of high-fuel-efficiency vehicles	3.1[c]

Source: Congressional Budget Office.

[a]Total cost between fiscal years 2009 and 2019, as estimated by the staff of the Joint Committee on Taxation.

[b] Total funding appropriated over the life of the program.

[c] Total net budgetary cost over the life of the program. Lawmakers originally appropriated $7.5 billion to cover the subsidy costs of loans made by the program. The Department of Energy (DOE) has obligated $3.5 billion of that budget authority on the basis of its initial estimates of the subsidy costs of the $8.4 billion in loans approved through May 2012 (including approximately $2.4 billion in loans identified as supporting the production of plug-in hybrid or all-electric vehicles). Of the $3.1 billion shown here, $1.6 billion is DOE's most recent revised estimate of the subsidy cost of the loans approved through May 2012, and the other $1.5 billion is CBO's projection of the subsidy costs that DOE will incur from the roughly $4 billion of the original $7.5 billion appropriation not yet obligated.

The Advanced Technology Vehicles Manufacturing (ATVM) program provides loans to U.S. automakers and parts manufacturers to help offset the cost of reequipping, expanding, or establishing plants to produce high-fuel-economy vehicles and their components. The $25 billion loan program was authorized by the Energy Independence and Security Act of 2007, and appropriations to cover the estimated $7.5 billion subsidy cost were provided by the Consolidated Security, Disaster Assistance, and Continuing

Appropriations Act, 2009. (The $25 billion is the total principal amount authorized; the $7.5 billion is the total estimated net cost of the loans to the government, after accounting for loan defaults, interest payments, and other aspects of the loan transactions.) As of May 2012, DOE had approved $8.4 billion in loans under the ATVM program.[7] A total of $3.5 billion was initially obligated to cover the estimated subsidy cost of those loans, although DOE has since revised that estimated cost downward to $1.6 billion.

However, only $2.4 billion of the loans approved so far are identified as supporting the production of electric vehicles (as opposed to other vehicle technologies). Of the $4.0 billion that DOE has not yet obligated from the $7.5 billion originally appropriated for subsidy costs, CBO anticipates that $1.5 billion will be spent in coming years.

Thus, the budgetary cost of the main federal programs that support electric vehicle technologies is estimated to total about $7.5 billion through 2019: $2.0 billion for the tax credits; $2.4 billion for grants to battery producers, intermediate suppliers, and demonstration projects; and $3.1 billion for the estimated subsidy cost of the loans provided by the ATVM program ($1.6 billion for loans already made and $1.5 billion for future loans).[8] Because the bulk of the ATVM loans were made to automakers for other purposes, the estimated budgetary cost of support for electric vehicles and related technologies through 2019 is $4.4 billion plus a portion of the $3.1 billion in ATVM subsidy costs—or a total of $5 billion to $6 billion.

In addition, the government now has or recently had some broader incentive programs (not shown in Table 1) for which producers or users of electric vehicles and related technologies are or were eligible. They include the following:

- The Advanced Energy Manufacturing Tax Credit, a tax credit equal to 30 percent of the cost of establishing a facility to develop advanced and clean energy technologies;
- The Alternative Fuel Infrastructure Tax Credit, a 30 percent tax credit that was available through 2011 to consumers and businesses that supported the installation of alternative fueling equipment;
- The Innovative Technology Loan Guarantee Program, an initiative that provides subsidized loans to promote technologies that reduce or sequester air pollutants or greenhouse gas emissions; and
- The Clean Cities Program, a partnership among the federal government, local governments, and private industries to reduce petroleum consumption in the transportation sector.

However, those four broader incentive programs are not major sources of support for electric vehicles or related manufacturing, because they have been used primarily for other purposes.

How the Tax Credits for Electric Vehicles Compare with Other Incentives

Of the federal government's current incentives for the use of electric vehicle technologies, the tax credits for the purchase of such vehicles have the most significant effect on vehicle ownership because they have the biggest impact on the price that consumers pay. The federal aid to producers of electric vehicles or components could lower the price of such vehicles, but much less than the tax credits do. For example, if production facilities are able to meet the schedule that DOE envisions for the Electric Drive Vehicle Battery and Component Manufacturing Initiative over the course of a 20-year plant life, and if the savings in production costs from that program's grants are fully passed on to consumers in the form of lower vehicle prices, the grants will reduce the cost of buying an electric vehicle by about $30 per kWh of battery capacity. For a 16 kWh electric vehicle, that savings would amount to about $500, much smaller than the $7,500 tax credit for purchasing a vehicle of that size. That comparison ignores other potential benefits of the grant program that could reduce vehicle prices over time—such as increased competition among automakers or more-rapid advances in technology because of increased research and development—but those effects are difficult to estimate and probably have a smaller impact on prices than the tax credits do. Other federal programs that aid producers of electric vehicles or components provide less assistance than does the Electric Drive Vehicle Battery and Component Manufacturing Initiative and thus will have even less effect on the retail price of electric vehicles.

Similarly, other types of programs that support electric vehicles appear to have had little effect to date on the market for such vehicles. Vehicle demonstration projects are aimed at helping to educate consumers about the viability of electric vehicles, and incentives to expand charging infrastructure are intended to make electric vehicles more convenient to own and less costly to operate. Those programs could have a more significant impact in future years, particularly if they stimulate consumer demand to the point that higher production volume greatly reduces the cost of electric vehicles.

USING FEDERAL TAX CREDITS TO PROMOTE SALES OF ELECTRIC VEHICLES

For electric vehicles to achieve the aims that supporters have for them—such as decreasing gasoline consumption, reducing emissions of greenhouse gases, and strengthening the U.S. automobile industry—consumers must buy those vehicles. Because of differences in vehicle design and technology, electric vehicles cost thousands of dollars more to purchase than conventional vehicles of comparable size and performance. At the same time, electric vehicles are less expensive to operate than other vehicles, because electricity is cheaper than gasoline per mile of travel. Thus, in deciding whether to buy an electric vehicle, consumers face a trade-off between its higher price and its lower operating cost compared with other types of vehicles.

The electric vehicle tax credits can influence that decision by offsetting some of the vehicles' higher purchase price. But how large would the credits need to be to make electric vehicles cost-competitive with other vehicles, in the sense of having similar total purchase and operating costs over the lifetime of a vehicle? The answer depends on the relative purchase and operating costs of different electric and nonelectric vehicles and on how consumers value the trade-off between current costs and future savings. By CBO's estimates, the credits are large enough for some electric vehicles (such as plug-in hybrids with small batteries) to be cost-competitive with conventional vehicles of the same size and performance, given current fuel prices. But for other electric vehicles (such as compact plug-in hybrids with relatively large batteries), the credits would have to be two or three times as large as they are now to make those vehicles cost-competitive.

Those results have several important caveats. First, they apply to electric vehicles at current prices. However, experts project that the prices of such vehicles will decline in coming years. In CBO's estimation, that decline may be significant enough that by the end of this decade, the tax credits, if still available, would make many electric vehicles less expensive to own and operate over the life of the vehicles than comparable nonelectric vehicles.

Second, although CBO's analysis focuses on the cost-competitiveness of electric vehicles, buyers also weigh other characteristics when deciding which vehicle to purchase. For example, electric vehicles may appeal to people who like to be "early adopters" of a new technology or who place great value on reducing their use of petroleum or their greenhouse gas emissions. Such buyers may purchase an electric vehicle even if the tax credits do not completely offset its higher lifetime costs. Other buyers—faced with uncertainty about whether electric vehicles will have a long enough range to satisfy their driving needs, whether opportunities to recharge the battery will be plentiful and recharging times will be short enough for convenience, or whether electric vehicle technologies will be sufficiently reliable compared with other technologies—may not purchase an electric vehicle even if the tax credits more than offset its higher lifetime costs. Either way, such noncost characteristics mean that electric and nonelectric vehicles are not perfect substitutes for one another, even if they are otherwise similar.

How the Tax Credits Affect the Relative Costs of Vehicle Ownership

For its analysis of the impact of the tax credits on vehicles' lifetime costs and on gasoline consumption and greenhouse gas emissions, CBO compared notional plug-in hybrid or all-electric vehicles with representative nonelectric vehicles in three different classes: low-fueleconomy trucks, average-fuel-economy light-duty vehicles, and high-fuel-economy compact cars.[9] That approach avoided the kind of apples-to-oranges comparison that would occur if CBO examined the costs and operating characteristics of a compact electric vehicle relative to those of a conventional light-duty truck.

The key results of that analysis are as follows:

- Given current prices for vehicles and fuel, in most cases the existing tax credits do not fully offset the higher lifetime costs of an electric vehicle compared with those of an equivalent conventional vehicle or traditional hybrid. For example, CBO estimates that a plug-in hybrid with a 16 kWh battery that is comparable in size and performance to an averagefuel-economy conventional vehicle (that is, one with a fuel economy of about 25 miles per gallon) would cost about $19,000 more to buy than the conventional vehicle. That plug-in hybrid would reduce the total discounted present value of fuel costs over an assumed 150,000-mile life by about $7,000 (based on average prices, in 2010 dollars, of $3.60 per gallon for gasoline and 12 cents per kWh for electricity and a discount rate of 10 percent), for a total difference in lifetime costs of about $12,000.[10] The $7,500 tax credit that applies to such a vehicle would need to be about 60 percent larger to make up that difference.

- Assuming that everything else is equal, the larger an electric vehicle's battery capacity, the greater its cost disadvantage relative to conventional vehicles—and thus the larger the tax credit needed to make it cost-competitive (see the top panel of Figure 1). The reason is that bigger batteries are more expensive, and the additional capacity provides value to drivers only on days when they are able to use that capacity before the battery is recharged again (which is assumed to occur once each day). Conversely, electric vehicles with small batteries are more cost-competitive. For example, the present-value difference in lifetime costs between a plug-in hybrid with a 4 kWh battery and a comparable average-fuel-economy conventional vehicle is about $2,400, CBO estimates—nearly the same as the current tax credit for that vehicle. But the difference in costs between a plug-in hybrid with a 16 kWh battery and a comparable average-fuel-economy conventional vehicle is about $12,000.
- All-electric vehicles are closer to being cost-competitive with conventional vehicles than are plug-in hybrids with the same size battery, but the tax credits would still need to be about 50 percent higher than they are now to fully offset the higher lifetime costs of an all-electric vehicle. The key reason is that the limits on how far all-electric vehicles can travel before needing to be recharged reduce their savings in fuel costs.[11] Without those limits—that is, if all-electric vehicles could be driven the same distance as conventional or plug-in hybrid vehicles—the $7,500 tax credit would generally be sufficient to make them cost-competitive.
- A larger tax credit is needed to make electric vehicles cost-competitive with higher-fuel-economy conventional vehicles. In the case of plug-in hybrid vehicles with 16 kWh batteries, for example, equalizing the lifetime costs of a highly fuel-efficient conventional compact car and a plug-in hybrid of comparable size and performance would require about $3,400 more in tax credits than equalizing the lifetime costs of a lowfuel-economy conventional light-duty truck and a comparable plug-in hybrid—that is, about $13,200 instead of $9,800 (see the the estimates for PHEV-16 in the top panel of Figure 1). The differences are smaller for all-electric vehicles. Because those vehicles will probably be driven for fewer total miles during a year, their overall lifetime cost is determined more by their purchase price than by future savings on fuel.

- The tax credits that would be needed to make plug-in hybrids cost-competitive are about the same whether those vehicles are compared with conventional vehicles or with traditional hybrids. The savings in fuel costs are smaller when the alternative to a plug-in hybrid is a traditional hybrid, but that reduction in savings is largely offset by a smaller difference in purchase price.
- Although tax credits are generally not large enough to ensure that electric vehicles are cost-competitive today, they may be sufficient to do so if they are available in later years.[12] Experts predict that prices for electric vehicles will decline in the years ahead as technological improvements and the effects of larger-scale production take hold. On the basis of those predictions, CBO projects that the gap in purchase price between an electric vehicle and a comparable conventional vehicle will narrow by an average of about 40 percent by 2020. At that point, the purchase price of a plug-in hybrid would exceed the price of a comparable conventional vehicle by about $700 per kWh of battery capacity, on average, compared with $1,200 per kWh now. With those vehicle prices, the tax credits—if still available—would generally be sufficient to make electric vehicles cost-competitive, taking into account the remaining difference in purchase prices and savings in fuel costs (see Figure 2).[13] Those projections incorporate the expectation that the fuel economy of all vehicles will rise over time, as will gasoline and electricity prices.

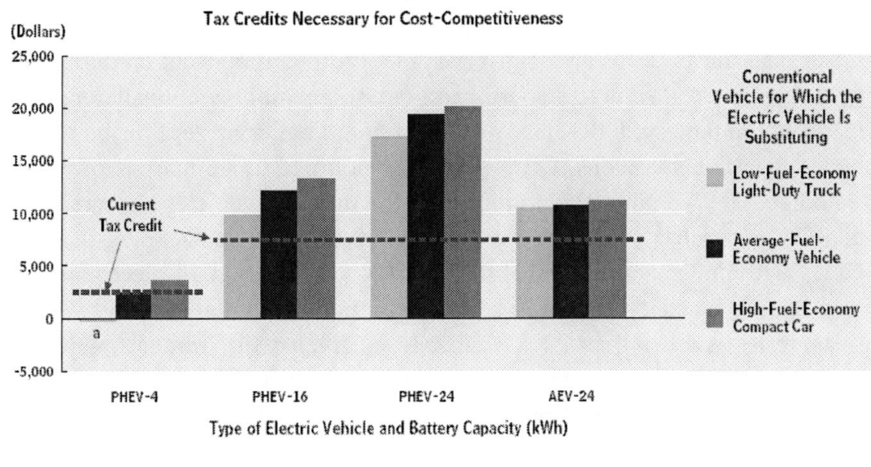

Figure 1. (Continued).

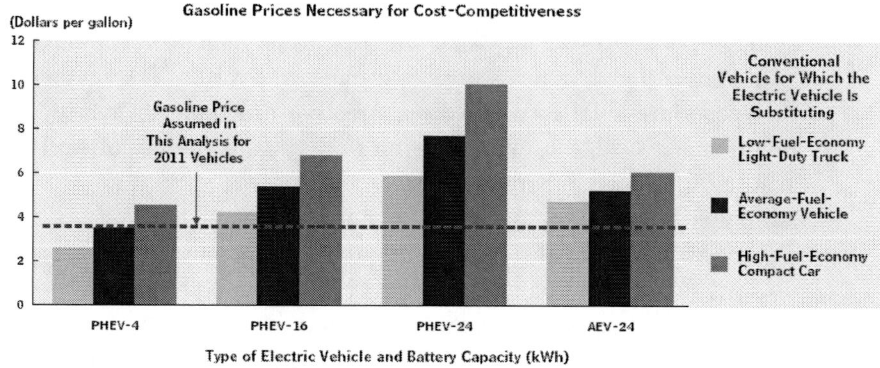

Source: Congressional Budget Office.

Notes: These results are based on an assumed discount rate of 10 percent and assumed prices (in 2010 dollars) of $3.60 per gallon for gasoline and 12 cents per kWh for electricity. Those and other assumptions are discussed in the appendix.

kWh = kilowatt-hours; PHEV-4 =plug-in hybrid electric vehicle with a 4 kWh battery; AEV-24 =all-electric vehicle with a 24 kWh battery.

a. Value of -$210. A negative value indicates that the discounted lifetime cost of an electric vehicle without the tax credit is less than the cost of an equivalent conventional vehicle.

Figure 1. Tax Credits and Gasoline Prices Necessary for Various Electric Vehicles to Be Cost-Competitive with Conventional Vehicles at 2011 Vehicle Prices.

The Basis for CBO's Cost Comparisons

Because the price of an electric vehicle, the maximum tax credit a buyer can receive, and the vehicle's energy and environmental benefits all depend on the size of the battery, CBO based its analysis on battery size. For the electric vehicles in its comparisons, CBO used four notional combinations of battery size and vehicle type intended to represent the range of vehicles expected to be available for purchase either now or at some point in the future: plug-in hybrid vehicles with battery capacities of 4, 16, or 24 kWh and an all-electric vehicle with a battery capacity of 24 kWh. CBO did not include all-electric vehicles with batteries smaller than 24 kWh because, given their limited range, such vehicles would probably not appeal to enough buyers for automakers to market them in significant numbers.

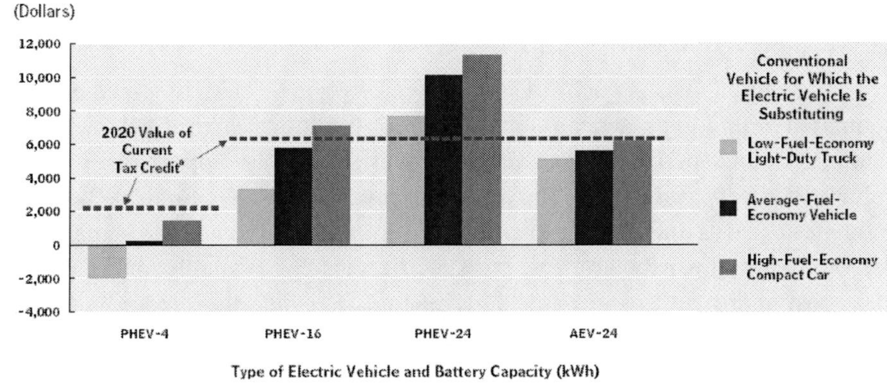

Source: Congressional Budget Office.

Notes: These results are based on an assumed discount rate of 10 percent and assumed prices (in 2010 dollars) of $3.90 per gallon for gasoline and 12 cents per kWh for electricity. Those and other assumptions are discussed in the appendix.

Vehicle prices in 2020 reflect a decline of approximately 40 percent in the difference between retail prices of electric vehicles and those of equivalent conventional vehicles compared with price differences in 2011.

A negative value for the necessary tax credit indicates that the discounted lifetime cost of an electric vehicle without the tax credit is less than the cost of an equivalent conventional vehicle.

kWh = kilowatt-hours; PHEV-4 =plug-in hybrid electric vehicle with a 4 kWh battery; AEV-24 =all-electric vehicle with a 24 kWh battery.

a. The amounts of the tax credit for electric vehicles are fixed in nominal terms. To calculate their value in 2020, CBO adjusted those amounts to reflect the expected growth of overall prices between now and 2020, which will reduce the inflation-adjusted value of the credit to $2,100 for a plug-in hybrid with a 4 kWh battery and $6,300 for all-electric vehicles and plug-in hybrids with batteries having a capacity of 16 kWh or more.

Figure 2. Tax Credits Necessary for Various Electric Vehicles to Be Cost-Competitive with Conventional Vehicles at 2020 Vehicle Prices.

Plug-in hybrids with 24 kWh batteries are also unlikely to play a significant role in the market. (They would appeal to relatively few buyers because they would be more expensive than plug-in hybrids with 16 kWh batteries, but drivers would benefit from the extra range provided by the larger battery only on days when their driving exceeded the range of a 16 kWh plug-in hybrid.) Nevertheless, CBO included those vehicles in its analysis to illustrate the cost characteristics and the energy and environmental benefits of driving a plug-in hybrid with a battery larger than 16 kWh (the smallest capacity that is eligible for a $7,500 credit) and to provide a direct comparison with a similarly sized all-electric vehicle.

In a given class, each of those four notional electric vehicles was compared with a hypothetical conventional or traditional hybrid vehicle of the same size and performance. Such comparisons only approximate actual vehicle choices available to buyers, for several reasons. First, CBO used estimates of costs and characteristics that were intended to be representative of the vehicles in a particular class, but specific vehicles available on the market may cost more or less to buy and operate. Second, the electric vehicles currently being produced do not have identical nonelectric counterparts. Third, electric vehicles do not perfectly substitute for otherwise identical nonelectric vehicles because differences in their driving range, ease of refueling, technology, and emissions are important to some buyers.

Although CBO compared electric vehicles with both conventional vehicles and traditional hybrids, the primary focus of this study is the trade-off between comparable electric and conventional vehicles. One reason for comparing electric vehicles with traditional hybrids is that potential buyers of electric vehicles are probably more likely to buy a traditional hybrid than a conventional vehicle, because they probably place greater weight on future savings in fuel costs or on the potential energy and environmental benefits of reducing gasoline consumption. Nevertheless, CBO focused on the comparison with conventional vehicles because sizable reductions in the gasoline use or greenhouse gas emissions of light-duty vehicles are possible only if conventional vehicles account for the bulk of such reductions—not only because traditional hybrids are fewer in number but also because they are already more fuel efficient than conventional vehicles. Consequently, to the extent that people who receive tax credits for buying electric vehicles would otherwise have bought traditional hybrids rather than conventional vehicles, the cost of the tax credits per unit of environmental gain is higher, as discussed later in this study.

Key Assumptions Underlying CBO's Analysis

To compare electric and nonelectric vehicles, CBO had to make numerous estimates and assumptions about such factors as the relative purchase prices of different types and sizes of electric and nonelectric vehicles, the amount of gasoline or electric power they use for each mile they travel, current and future prices for gasoline and electricity, and the rate at which consumers discount the value of future savings relative to current costs. Those estimates and assumptions are described in detail in the appendix, but several key ones are discussed below.

Vehicle Prices

On the basis of estimates by other analysts, CBO concludes that, on average, the difference in purchase price between a plug-in hybrid vehicle and a conventional vehicle of similar size and performance consists of a fixed component of about $4,000 (which reflects cost differences that are independent of the size of the electric vehicle's battery) and a variable component of about $950 per kWh (which reflects costs that depend directly or indirectly on the size of the battery).[14] On that basis, a 16 kWh plug-in hybrid can be expected to cost about $19,000 more to buy than a comparable conventional vehicle [$4,000 + ($950 x 16)]—or an average price difference of about $1,200 per kWh of battery capacity.

All-electric vehicles cost less to buy than plug-in hybrids with the same battery capacity because they do not require certain components (such as an internal combustion engine) or they use simpler systems (such as transmissions). CBO estimates that the difference in purchase price between a 24 kWh all-electric vehicle and a comparable conventional vehicle is about 40 percent smaller than the price difference for a plug-in hybrid with that size battery. Thus, a 24 kWh all-electric vehicle is estimated to cost about $16,000 more than a comparable conventional vehicle.[15]

For simplicity, CBO assumes that the availability of tax credits does not increase the purchase price of electric vehicles. In practice, the tax credits are likely to spur demand for electric vehicles enough to make prices for those vehicles higher than they would have been otherwise, thereby offsetting a portion of the tax credit provided. The extent to which that happens will depend on whether manufacturers or buyers of electric vehicles are more sensitive to changes in prices; little evidence exists on which to make that judgment.

Fuel Prices

For its analysis of current vehicles, CBO assumes real (inflation-adjusted) prices of $3.60 per gallon for gasoline and 12 cents per kWh for electricity. Those amounts, which are in 2010 dollars, are based on recent projections by the Energy Information Administration, averaged over the years that current vehicles will be in use.[16]

Discount Rate for Future Savings

CBO assumes that people discount the value of future savings in fuel costs at a rate of 10 percent per year, a value consistent with research about how consumers discount future savings in other types of energy costs.[17]

Value of the Tax Credit

CBO's comparisons also incorporate the assumption that buyers of new vehicles receive the full value of the tax credit for which they are eligible. Because the credits are not refundable, however, people with a small income tax liability may be eligible to receive only a fraction of the credit available to them. (How the current credits compare with refundable tax credits or rebates is discussed near the end of this report.) In practice, however, most purchases of new vehicles—especially fairly expensive vehicles—are made by people in higher-income households, who are more likely to have enough federal income tax liability to apply the full value of the credit.

Effects of Differing Assumptions About Fuel Prices and Discount Rates

Whether electric vehicles are cost-competitive with conventional vehicles depends in part on the prevailing prices for gasoline and electricity and on how people compare the value of savings in future years with the higher cost of purchasing an electric vehicle today. CBO performed additional analyses to gauge the impact of those two factors.

Fuel Prices

Lower electricity prices or higher gasoline prices would reduce the relative cost of owning electric vehicles and thereby reduce the size of the tax credit necessary to make electric vehicles cost-competitive. Of those two types of fuel prices, gasoline prices have more potential to narrow the cost gap. With gasoline prices of $6 a gallon, for example, the lifetime costs of many types of electric vehicles would be less than or equal to the costs of conventional vehicles, given the current tax credits (see the bottom panel of Figure 1 on page 8). But even if electricity were free, the tax credits would still need to be about twice as high as the current ones, in many cases, before electric vehicles would be cost-competitive.

In general, the gasoline price necessary to equalize costs is lower when the alternative to an electric vehicle is a low-fuel-economy conventional vehicle than when it is a high-fuel-economy conventional vehicle. For instance, with the current tax credits, equalizing the costs of a conventional light-duty truck and an equivalent electric vehicle would require a gasoline price of no more than $6 per gallon, whereas equalizing the costs of a conventional compact car and an equivalent electric vehicle would require a price as high as $10 a gallon (in the case of a plug-in hybrid with a 24 kWh battery). Plug-in hybrids with smaller batteries can be cost-competitive (with the tax credits) when gasoline prices are in the $3 to $5 per gallon range. When the alternative to an electric vehicle is a traditional hybrid (not shown in Figure 1), those break-even gasoline prices are generally about $1 to $2 per gallon higher.

Discount Rate for Future Savings

People who place less value on future savings in fuel costs and who focus more on the purchase price of a vehicle require larger tax credits or higher gasoline prices to consider an electric vehicle cost-competitive. CBO generally assumed a discount rate of 10 percent, but because that assumption has a significant impact on estimates of the present value of lifetime costs, CBO also assessed how discount rates half or twice that amount would affect the cost-competitiveness of electric vehicles (see Figure 3). The higher the discount rate, the less value is attributed to the savings in fuel costs from acquiring an electric vehicle and the larger the tax credit that is needed to make such a vehicle costcompetitive.[18] Thus, compared with people who discount future costs or savings at a rate of 5 percent a year, people who use a discount rate of 20 percent would require a tax credit about 50 percent larger (roughly $15,000 instead of $10,000) or gasoline prices about twice as high (approximately $8 rather than $4) to consider the lifetime costs of an average 16 kWh plug-in hybrid vehicle equal to those of a comparable conventional vehicle. Analyzing the effects of discount rates of 5 percent, 10 percent, and 20 percent yields similar results when the alternative to an electric vehicle is a traditional hybrid (not shown in Figure 3) rather than a conventional vehicle.

USING FEDERAL TAX CREDITS FOR ELECTRIC VEHICLES TO ADDRESS ENERGY AND ENVIRONMENTAL GOALS

The transportation sector accounts for about 70 percent of petroleum use in the United States and about 35 percent of the nation's carbon dioxide emissions attributable to human activity. Because nearly all of the energy used in the transportation sector is petroleum based, reducing emissions from that sector would require cutting petroleum consumption through some combination of improving the fuel economy of vehicles, substituting nonpetroleum fuels for petroleum-based ones, or traveling less.

The current tax credits for electric vehicles are one tool that policymakers have adopted to pursue the goals of decreasing petroleum consumption and greenhouse gas emissions in the transportation sector. Electric vehicles use no gasoline when running on electric power, and in the case of plug-in hybrids and traditional hybrids, they use about one-third less fuel when running on gasoline power than do conventional vehicles of similar size and performance. Driving an electric vehicle instead of a conventional vehicle can also significantly lessen greenhouse gas emissions, depending on the type of vehicle used and the emissions released in generating electricity for it.

CBO's analysis suggests two conclusions about the effectiveness of the tax credits for electric vehicles in advancing those energy and environmental goals:

- In the short term, the tax credits are likely to have little or no impact on total gasoline consumption and greenhouse gas emissions.
- In the long term, the credits might decrease gasoline use and emissions, but how cost-effectively they would do so is unknown.

CBO reached those conclusions by considering the interactions between the tax credits and the federal government's corporate average fuel economy standards for cars and light-duty trucks (for more details about CAFE standards, see Box 2). The tax credits have the direct effect of increasing sales of electric vehicles. The credits also have the indirect effect of boosting sales of high-fueleconomy conventional vehicles and traditional hybrids, by encouraging sellers of such vehicles to lower their prices to better compete with electric vehicles. However, as automakers seek to comply with the current CAFE standards, they are expected to produce a mix of vehicles that, on average, meets the CAFE standards but does not significantly exceed them. Consequently, the more electric and other high-fuel-economy vehicles that are sold because of the tax credits, the more low-fuel-economy vehicles that automakers can sell and still meet the standards.

With CAFE standards in place, therefore, putting more electric (or other high-fuel-economy) vehicles on the road will produce little or no net reduction in total gasoline consumption and greenhouse gas emissions. Hence, CBO expects that the tax credits for current and future sales of electric vehicles will have little net effect through 2021, the period covered by the most recently finalized CAFE standards.

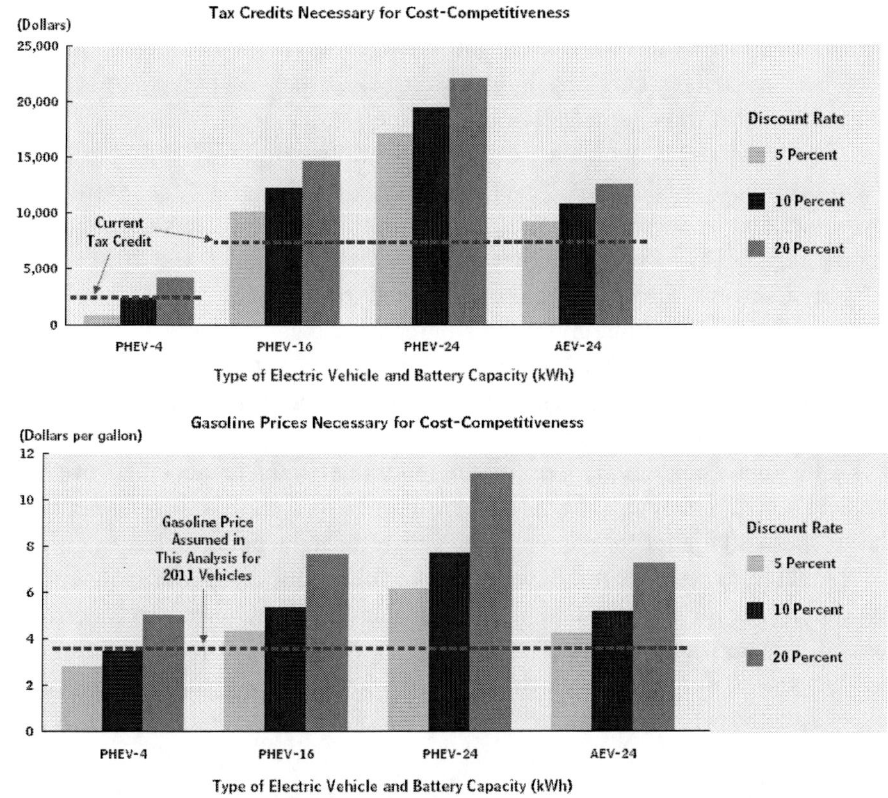

Source: Congressional Budget Office.

Notes: These results are based on the assumptions that electric vehicles are substituting for average-fuel-economy conventional vehicles and that prices (in 2010 dollars) are $3.60 per gallon for gasoline and 12 cents per kWh for electricity. Those and other assumptions are discussed in the appendix.

kWh = kilowatt-hours; PHEV-4 =plug-in hybrid electric vehicle with a 4 kWh battery; AEV-24 =all-electric vehicle with a 24 kWh battery.

Figure 3. Tax Credits and Gasoline Prices Necessary for Various Electric Vehicles to Be Cost-Competitive at 2011 Vehicle Prices, Using Different Discount Rates.

Box 2.
Corporate Average Fuel Economy Standards

Since 1975, the federal government's corporate average fuel economy (CAFE) standards have specified an average minimum number of miles that new passenger cars and light-duty trucks (such as pickup trucks or minivans) must travel on a gallon of gasoline.

Those standards, which are set by the National Highway Traffic Safety Administration (NHTSA), were largely unchanged for vehicles manufactured between 1990 and 2010. However, NHTSA has set steadily increasing CAFE standards for new vehicles through model year 2021.

In addition, the Environmental Protection Agency (EPA) has set parallel standards to limit emissions of greenhouse gases from new vehicles (measured in grams of emissions released per mile of travel) through model year 2025.[1] Those standards are expected to be met largely through improvements in fuel efficiency.

The fuel economy target for an individual vehicle depends on the vehicle's size and type: The targets are higher for small vehicles than for large ones and higher for passenger cars than for light-duty trucks.

An automaker does not have to meet a fuel economy standard for each type and size of vehicle that it sells. Instead, it is required to meet an overall standard that depends on the mix of vehicles that it sells during a model year. To comply with that overall standard, the manufacturer can exceed the targeted improvements in fuel economy for some vehicles sold to make up for shortfalls in fuel economy for other vehicles sold.

Under the standards in effect for model years 2012 to 2016, the total average fuel economy of new light-duty vehicles sold by all manufacturers is scheduled to rise from 29.7 miles per gallon (mpg) this year to 34.1 mpg in 2016.

Under the standards for 2017 to 2025, NHTSA and EPA expect average fuel economy for new vehicles to reach about 40 mpg by 2021 and 50 mpg by 2025. (Those figures are estimates based on projections of the number and type of new vehicles that will be sold in a given year.)

Complying with rising CAFE standards is likely to be costly for automakers. Thus, manufacturers who intend to comply with the rules (rather than pay a fine) are likely to produce vehicles that just meet the standards without significantly exceeding them.

Automakers have some flexibility in meeting CAFE standards; that flexibility serves to shift the burden of compliance from one year to another or from one producer to another, but it does not reduce the burden overall.[2]

1. NHTSA and EPA recently finished formulating standards for 2017 and later years. Because NHTSA has authority to set CAFE standards for only five years at a time, its rules are binding (required of automakers) only through model year 2021. A separate rulemaking will be necessary to set CAFE standards for 2022 through 2025. NHTSA has specified nonbinding fuel economy standards for those four years to help automakers with their product planning and to harmonize with EPA's emissions standards, which apply through 2025. See National Highway Traffic Safety Administration, "NHTSA and EPA Set Standards to Improve Fuel Economy and Reduce Greenhouse Gases for Passenger Cars and Light Trucks for Model Years 2017 and Beyond," www.nhtsa.gov/staticfiles/rulemaking/pdf/cafe/CAFE_2017 - 25_Fact_Sheet.pdf
2. To help meet its target for a given year, an automaker can apply credits that it earned by exceeding its target sometime in the previous five years or that it purchased from another producer. It can also achieve compliance up to three years after the fact through the retroactive use of credits earned later.

The tax credits for current vehicle sales could decrease gasoline use and emissions in later years, however, if sales of electric vehicles lead policymakers to set CAFE standards for 2022 and beyond at higher levels than they would otherwise. Indeed, as discussed below, there is evidence that regulators considered expectations about the future prevalence of plug-in electric vehicles when they set binding CAFE standards for model years 2020 and 2021 and preliminary standards for 2022 through 2025. To the extent that regulators' expectations were informed by past sales of electric vehicles, the tax credits provided for those sales are likely to have an effect on gasoline consumption and greenhouse gas emissions starting in 2020.

How long the tax credits might influence fuel economy standards depends in part on the success of the electric vehicle industry. If the credits help the industry achieve commercial viability and capture a significant share of the vehicle market, they might continue to influence CAFE standards—and thereby reduce gasoline consumption and greenhouse gas emissions—for many years after they have been phased out. Conversely, if electric vehicles never achieve commercial success, the influence of the tax credits on CAFE standards will be small or short-lived, and the credits will prove not to have been cost-effective in either the short or the long term.

Direct Effects of the Tax Credits in the Short Run

In CBO's analysis, the direct effects of the tax credits are the effects that follow from additional sales of electric vehicles. The indirect effects, by contrast, follow from the credits' impact on sales of other types of vehicles.

To measure the direct effects of the tax credits relative to energy and environmental goals, CBO used an average cost measure—referred to as the government's cost—that compares the value of the credits with the reductions in gasoline consumption and greenhouse gas emissions that result from driving electric vehicles in place of other vehicles of similar size and performance. That cost is the payment—per gallon of gasoline saved or metric ton of carbon dioxide equivalent (CO_2e) emissions reduced— that the federal government effectively makes to buyers of new electric vehicles to bring about those reductions (before incorporating the indirect effects of the credits on the mix of other new vehicles sold).[19] By CBO's estimate, those costs of reducing gasoline use and emissions by substituting electric vehicles for comparable conventional vehicles or traditional hybrids are relatively high compared with the costs of other policies aimed at lowering gasoline consumption or emissions. (Those comparisons are discussed in detail later in this report.)

Measuring Direct Cost-Effectiveness

CBO separately estimated the government's cost of using the tax credits to decrease gasoline consumption or to reduce greenhouse gas emissions. In fact, the credits serve both purposes simultaneously—as well as others, such as improving air quality or supporting the development of U.S. industries. In a full benefit-cost analysis, each of those benefits would be translated into dollar amounts, and the total dollar value of the benefits would be compared with the costs. Nevertheless, the cost-effectiveness analysis described here, which treats reducing gasoline consumption and greenhouse gas emissions as if each were the only policy goal, is sufficient to allow comparisons between the tax credits and some other policy tools that address similar goals.[20]

CBO's estimates of the cost-effectiveness of the tax credits account for the likelihood that some of the credits will be provided for vehicle purchases that would have been made even without them. In such cases, although the tax credits reduce the cost of an electric vehicle, they do not result in any energy or environmental benefits. On the basis of research about how sales of traditional hybrids have responded to federal and state incentives, CBO estimates that about 30 percent of current and future sales of electric vehicles will be attributable to the tax credits, and 70 percent would have occurred even without the credits (see the appendix). The implication is that only about one-third of the credits will produce energy or environmental benefits, which means that the cost to the federal government of those benefits will be about three times higher than it would be if the tax credits were responsible for all electric vehicle sales.

This analysis does not consider other factors that could influence the cost-effectiveness of the tax credits. One such factor is the possibility that owning an electric vehicle might cause people to change how much they drive. On the one hand, the lower per-mile cost of driving an electric vehicle could lead to an increase in travel, eliminating some of the reductions in gasoline consumption and greenhouse gas emissions that would otherwise occur and thereby increasing the government's cost of achieving those reductions. On the other hand, if worries about exhausting the charge in their battery caused owners of electric vehicles to drive less, the reductions in gasoline consumption and greenhouse gas emissions would be greater and the cost of achieving those reductions would be lower.[21]

The Cost to the Government of Direct Reductions in Gasoline Consumption

The direct effects of the tax credits on gasoline use depend on many of the same factors that affect the relative cost of owning an electric vehicle: the fuel economy of the vehicle that a consumer would otherwise buy in the absence of an electric vehicle, the type of electric vehicle (plug-in hybrid or all-electric), and the capacity of its battery (which determines the size of the available tax credit).

If the alternative to an electric vehicle is another high-fuel-economy vehicle, the reductions in gasoline consumption will be relatively costly (everything else being equal) because the more fuel efficient the alternative vehicle is, the smaller the gain from buying an electric vehicle instead. For example, the tax credits cost the government about $4 per gallon of gasoline saved if someone drives a 16 kWh plug-in hybrid light-duty truck instead of a comparable conventional light-duty truck that gets about 15 miles to the gallon (see the top panel of Figure 4).

That cost is twice as high when comparing a 16 kWh plug-in hybrid with an otherwise similar highfuel-economy conventional compact car that gets nearly 40 miles to the gallon. The cost per gallon saved will generally be even greater when comparing an electric vehicle with a traditional hybrid, owing to the latter's high fuel economy.

For example, that cost is about $14 per gallon of gasoline saved when a 16 kWh plug-in hybrid is driven instead of a comparable high-fuel-economy traditional hybrid (not shown in Figure 4).

Plug-in hybrids decrease gasoline consumption at a lower cost to the federal government than do all-electric vehicles with the same battery capacity. For instance, compared with an average-fuel-economy conventional vehicle, the cost per gallon of gasoline saved is about 30 percent less with a 24 kWh plug-in hybrid than with a 24 kWh all-electric vehicle (about $5 versus $7).

The reason is that all-electric vehicles will probably be driven for fewer miles in a year (because of their smaller range), so the reduction in gasoline consumption from driving those vehicles in place of conventional vehicles will be lower.

For a given type and size of electric vehicle, the cost per gallon of reducing gasoline consumption through the tax credits increases as the vehicle's battery capacity rises from 4 kWh to 16 kWh.

Larger capacity allows electric vehicles to be driven farther on electricity, but the additional gallons saved may be small relative to the additional tax credit provided. (On days when little driving occurs, for example, small- and large-battery electric vehicles reduce gasoline use by about the same amount, although the latter are eligible for a larger tax credit.) Above 16 kWh, additional increases in battery capacity decrease the government's cost of reducing gasoline consumption: The tax credit remains at the maximum value of $7,500, but the driving range on electricity—and the attendant reduction in gasoline use—continues to grow.

The Cost to the Government of Direct Reductions in Greenhouse Gas Emissions

The direct effects of the tax credits on emissions of greenhouse gases depend in part on the same factors that affect the cost of reducing gasoline consumption, because each gallon of gasoline not used yields a corresponding decrease in emissions. Thus, the greater the fuel efficiency of the vehicle for which an electric vehicle is substituting, the higher the cost per ton of emissions reduced.

That cost is also higher for all-electric vehicles than for plug-in hybrids, and for large-battery electric vehicles than for small-battery ones.

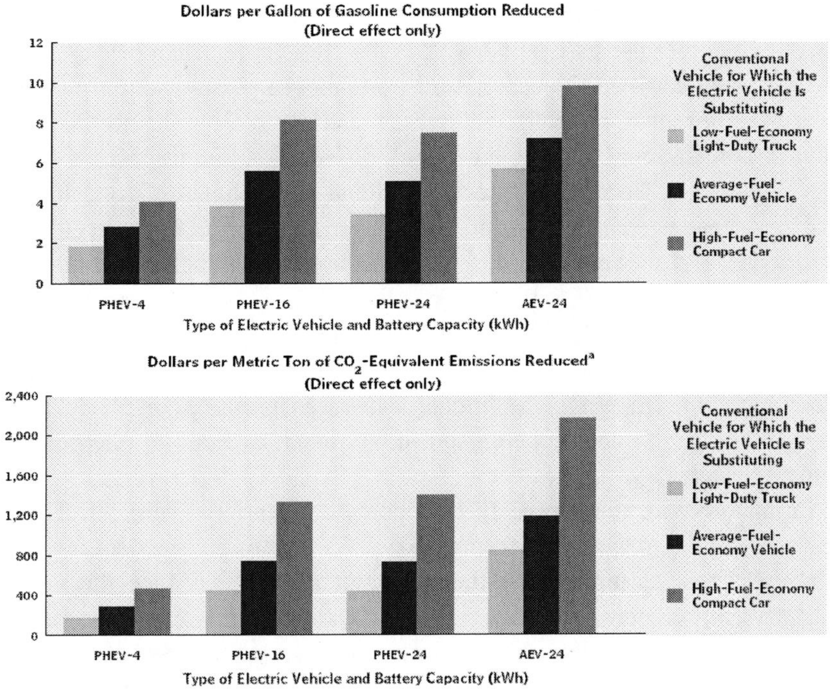

Source: Congressional Budget Office.

Notes: These results are based on the assumption that fuel with average emissions is used to produce electricity, as well as on the following assumptions: 2011 vehicle prices, current tax credits, a discount rate of 10 percent, and prices (in 2010 dollars) of $3.60 per gallon for gasoline and 12 cents per kWh for electricity. Those and other assumptions are discussed in the appendix.

The results do not reflect the indirect effects of electric vehicle sales on the mix of other vehicles sold.

Because individual greenhouse gases vary in their warming characteristics and persistence in the atmosphere, researchers commonly measure emissions in kilograms or metric tons of carbon dioxide equivalent-the amount of carbon dioxide that would cause an equivalent amount of warming over 100 years.

kWh = kilowatt-hours; PHEV-4 =plug-in hybrid electric vehicle with a 4 kWh battery; AEV-24 =all-electric vehicle with a 24 kWh battery; CO_2 = carbon dioxide.

a. Because individual greenhouse gases vary in their warming characteristics and persistence in the atmosphere, researchers commonly measure emissions in kilograms or metric tons of carbon dioxide equivalent—the amount of carbon dioxide that would cause an equivalent amount of warming over 100 years.

Figure 4. Cost to the Federal Government of Using Electric Vehicle Tax Credits to Reduce Gasoline Consumption and Greenhouse Gas Emissions.

The cost of direct reductions in greenhouse gas emissions also depends on the emissions that are associated with a vehicle's use but not released during that use. The electric utility sector emits greenhouse gases when producing the power used to recharge electric vehicles' batteries, to manufacture all types of vehicles (conventional, traditional hybrid, or electric), and to process the fuels that vehicles use. Because of those additional components of total life-cycle emissions (that is, of the emissions released directly or indirectly for each vehicle driven), the contrast between electric and nonelectric vehicles is smaller for greenhouse gas emissions than for gasoline consumption. In general, electric vehicles have lower life-cycle emissions than other vehicles do, because the amount of emissions saved by using less gasoline is generally greater than the amount of additional emissions produced in the electricity sector. However, much depends on how much carbon is released when producing electricity.

Compared with an average-fuel-economy conventional vehicle, an electric vehicle of similar size and performance will have about 35 percent lower life-cycle emissions, assuming that the power plants that produce the vehicle's electricity emit greenhouse gases at a rate equal to the national average for the electricity sector. Under those circumstances, compared with such a conventional vehicle, an electric vehicle will be responsible for only about half the amount of emissions per mile traveled when running on electric power and (in the case of plug-in hybrids) about two-thirds the amount of emissions per mile when running on gasoline power. The cost to the government of using the electric vehicle tax credits to achieve those reductions ranges from \$300 to \$1,200 per metric ton of CO_2e emissions reduced, depending on the battery size of the electric vehicle that is substituting for an average-fueleconomy conventional vehicle (see the darkest bars in the bottom panel of Figure 4). Smaller-battery vehicles have lower costs to the government, primarily because they are eligible for smaller tax credits.

Different assumptions about the amount of carbon released when producing electricity or about the type of vehicle for which an electric vehicle is substituting lead to a range of estimates for each type of electric vehicle and battery size. For example, the government's cost of reducing emissions can be much higher when electric power comes from coal-burning plants. Coal releases more greenhouse gas emissions when burned than any other fossil fuel—roughly twice the national average for the electricity sector. With such high-emissions electricity, the cost per metric ton of CO_2e emissions reduced when an electric vehicle substitutes for an average-fueleconomy conventional vehicle ranges from \$350 to \$4,400 (see Figure 5). All-electric vehicles account for the upper end of that range: Because they are unlikely to be driven as extensively as other vehicles, they produce smaller reductions in emissions, and consequently, the government's cost for those reductions is higher.

Conversely, the cost of reducing emissions can be far lower when electric power is produced from low-carbon sources, such as nuclear and hydroelectric power. In that case, the government's cost per metric ton of CO_2e emissions reduced when an electric vehicle substitutes for an average conventional vehicle ranges from about $230 to $630. That cost would be as low as $150 per metric ton if an electric vehicle with a small battery substituted for a low-fuel-economy light-duty truck (not shown in Figure 5).

The reduction in greenhouse gas emissions is lowest, and the cost per metric ton is highest, when electric vehicles substitute for traditional hybrids—possibly to the point where driving an electric vehicle instead of a traditional hybrid can cause life-cycle emissions to increase. That can happen when an electric vehicle replaces a traditional hybrid in an area where electricity comes mainly from coal-fired generation. In that case, the emissions from coal-fired power plants may be high enough (and the amount of gasoline used by a traditional hybrid low enough) that more emissions are released per mile traveled on electric power than per mile traveled on gasoline. In such a case, the tax credits effectively subsidize the release of additional greenhouse gas emissions.

Indirect Effects of the Tax Credits in the Short Run

Besides directly affecting energy use and emissions through their impact on the number of electric vehicles sold, the tax credits indirectly affect energy use and emissions in the near term by influencing the total average fuel economy of new nonelectric vehicles sold. Those indirect effects take two forms, which work in opposite directions. Both effects reduce sales of nonelectric vehicles that are moderately fuel efficient, but one increases sales of high-fuel-economy vehicles and the other increases sales of low-fuel-economy vehicles.

One effect works through price competition. Providing tax credits for purchases of new electric vehicles lowers their cost relative to that of other vehicles of comparable size and performance. To compete, automakers are likely to reduce the price of other high-fuel-economy vehicles to maintain their sales. A study of the tax credits that were available for traditional hybrids until 2011 estimated that their indirect effect on the average fuel economy of nonhybrids was about one and a half times the direct effect of driving those hybrids in place of conventional vehicles.[22] For reasons discussed in the appendix, CBO expects the price-competition effect to be smaller for the electric vehicle tax credits: only about half the size of their direct impact on average fuel economy.

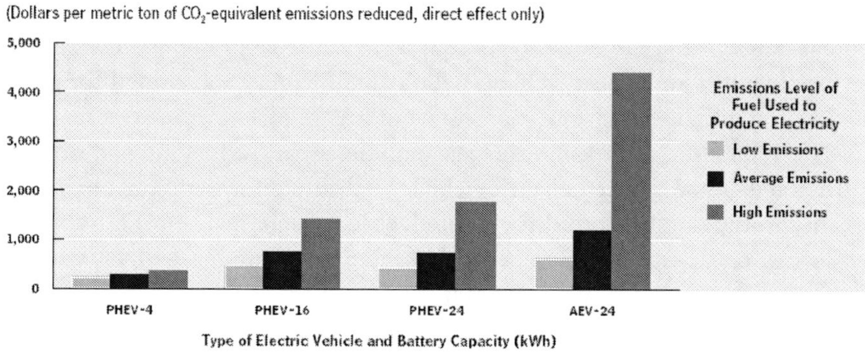

Source: Congressional Budget Office.

Notes: These results are based on the assumption that electric vehicles are substituting for average-fuel-economy conventional vehicles, as well as on the following assumptions: 2011 vehicle prices, current tax credits, a discount rate of 10 percent, and prices (in 2010 dollars) of $3.60 per gallon for gasoline and 12 cents per kWh for electricity. Those and other assumptions are discussed in the appendix.

The results do not reflect the indirect effects of electric vehicle sales on the mix of other vehicles sold.

"Low emissions" represents electricity generation from nuclear power or renewable energy sources; "average emissions" represents the current average for the electric utility sector (which is close to the typical emissions level of electricity generation from natural gas); "high emissions" represents electricity generation from coal.

Because individual greenhouse gases vary in their warming characteristics and persistence in the atmosphere, researchers commonly measure emissions in kilograms or metric tons of carbon dioxide equivalent—the amount of carbon dioxide that would cause an equivalent amount of warming over 100 years.

CO_2 = carbon dioxide; kWh = kilowatt-hours; PHEV-4 =plug-in hybrid electric vehicle with a 4 kWh battery; AEV-24 = all-electric vehicle with a 24 kWh battery.

Figure 5. Cost to the Federal Government of Using Electric Vehicle Tax Credits to Reduce Greenhouse Gas Emissions When Electricity Is Produced Using Fuels with Different Carbon Intensities.

The dominant indirect effect, however, is a consequence of federal CAFE standards. Regulators recently revised those standards, extending them through

2021 (and the parallel standards for greenhouse gas emissions through 2025; see Box 2 on page 14). The CAFE rules call for average fuel efficiency for all new light-duty vehicles to reach about 40 miles per gallon in 2021, and the greenhouse gas rules imply average fuel efficiency of about 50 miles per gallon in 2025. Those increases will probably be costly for automakers to meet, so vehicle manufacturers are likely to upgrade the average fuel economy of their fleets only to the point where it just meets the standards. In doing so, manufacturers may seek additional improvements in fuel efficiency for certain vehicle models to offset shortfalls in other models or to make it easier to meet further increases in the standards in later years. To the extent that the tax credits increase sales of electric vehicles and, through the price-competition effect, sales of nonelectric vehicles with high fuel efficiency, they also allow automakers to pursue fewer improvements in fuel efficiency in other vehicles and to sell a larger number of low-fuel-economy vehicles.[23] As a result, the credits have little net impact overall on energy use and emissions while a given set of CAFE standards remains in effect.[24]

Effects of the Tax Credits in the Long Run

When policymakers revise CAFE standards, the tax credits can have an impact on gasoline consumption and greenhouse gas emissions in the long run by influencing those revisions. In particular, if the tax credits cause regulators to expect higher sales of electric vehicles than would otherwise be the case, regulators may choose to set higher CAFE standards than they would otherwise—thus raising the average fuel economy of all vehicles sold in the years covered by those standards and thereby reducing gasoline consumption and greenhouse gas emissions. Some evidence suggests that the sales of electric vehicles that have already occurred will have an impact on gasoline consumption and emissions starting in model year 2020. In proposing the latest round of CAFE standards, the National Highway Transportation Safety Administration (NHTSA) indicated that it took into account the expected role of plug-in hybrid vehicles in 2020 and later years.[25] Presumably, NHTSA's expectations were influenced not only by input from automakers and other stakeholders about the future availability and technical capability of electric vehicles and other alternative vehicle technologies but also by observed vehicle sales to date.[26] Because the extent of that influence is unknown, however, CBO cannot evaluate the long-term cost-effectiveness of the electric vehicle tax credits. Also unknown is the length of time during which the credits will continue to have an impact. They could have energy and environmental benefits extending many years into the future if they help give the electric vehicle industry enough time to develop to the point that significant demand for such vehicles persists for the long term. Indeed, helping the U.S. electric vehicle industry reach commercial viability may be another goal of policy-makers, in addition to energy and environmental goals. Whether the tax credits and other federal incentives for electric vehicles discussed in this study will have such lasting effects is uncertain. Three alternative paths illustrate the possibilities.

Possibility 1: Electric Vehicles Fail to Achieve Significant Consumer Acceptance

Federal incentives will clearly not have achieved the goal of helping to bring about the widespread use of electric vehicles if those vehicles never attain a significant share of the U.S. automobile market. Traditional hybrids, for example, have been available for about 10 years and were eligible for federal tax credits for much of that time, but they currently account for less than 3 percent of new-vehicle sales. Unless consumers view electric vehicles very differently in the years ahead, sales of electric vehicles will probably be no greater than those of traditional hybrids because of their higher purchase price and lengthy recharging times.

Electric vehicles will also face competition from other types of vehicles. For the most part, future vehicles—both conventional vehicles and traditional hybrids—will be more fuel efficient than their current counterparts. In addition, other types of alternative vehicles may become available in coming years, such as vehicles that run on hydrogen fuel cells or compressed natural gas.

Possibility 2: The Electric Vehicle Industry Becomes Self-Sustaining in the Near Future

Even if electric vehicles attain a significant market share, it might not be because of the effects of federal incentives. That could happen if the life-cycle costs of electric vehicles become comparable with those of conventional or traditional hybrid vehicles in the next several years and would have done so even without the federal incentives. Although unlikely, such an outcome could result from a combination of dramatic increases in gasoline prices and technological breakthroughs that significantly reduce the costs of electric vehicles. In that case, the primary effect of the incentives would have been to boost short- and medium-term sales of electric vehicles, with no significant impact on long-term sales.

Possibility 3: Incentives for Electric Vehicles Promote Sales Until the Industry Becomes Self-Sustaining

Federal incentives could have a lasting impact on sales of electric vehicles—as well as on petroleum use and greenhouse gas emissions—if they are able to promote sales long enough for gradual technological progress to reduce the life-cycle costs of electric vehicles to levels comparable with those of other vehicles. Prices for electric vehicles could decline enough by the end of the decade to make many types cost-competitive with other vehicle options if the tax credits remained at their current levels. Even then, however, the continued possibility that electric vehicles could not be driven as far or refueled as quickly as conventional or traditional hybrid vehicles might continue to limit their sales. The federal incentives could also have a lasting impact if they help overcome barriers to the use of electric vehicles that could not be overcome otherwise by technological advances or market developments. For example, the benefit of owning an electric vehicle might depend on the number of such vehicles in use. Low sales would provide little incentive for the development of a commercial-grade, high-speed recharging infrastructure, and in turn, the lack of a recharging infrastructure away from home would limit additional vehicle sales. Supporting sales of 200,000 or more electric vehicles per manufacturer— thereby making it more likely that the industry will eventually reach commercial viability—could encourage the development of recharging and other technologies that would eventually help promote future vehicle sales by themselves.

COMPARING THE TAX CREDITS WITH OTHER RECENT SUBSIDY PROGRAMS IN THE TRANSPORTATION SECTOR

Besides the current tax credits for the purchase of new electric vehicles, some other recent subsidy policies have focused on the automobile industry:

- Tax credits for companies that blend biofuels with petroleum fuels, two of which (the credits for corn ethanol and biodiesel) expired in 2011 and the other of which (the credit for cellulosic ethanol) will expire at the end of 2012;

- The 2009 "Cash for Clunkers" program, in which buyers of new vehicles received direct subsidies of up to $4,500 when trading in an eligible lower-fueleconomy vehicle for a higher-fuel-economy vehicle; and
- Tax credits of up to $3,400 for the purchase of new traditional hybrid vehicles, which expired in 2010.

Only the biofuel tax credits and the "Cash for Clunkers" program had the potential to reduce gasoline consumption in the short run. Like the electric vehicle tax credits, the tax credits for traditional hybrids could affect fuel use and emissions only through their impact on the level of future CAFE standards.[27] In comparisons of the overall cost-effectiveness of those subsidy programs, short-term effects are only part of the story. The long-term effects of the tax credits for electric vehicles depend on factors such as their impact on policy-makers' expectations about future sales of electric vehicles and the extent to which those sales could help auto-makers meet future changes in CAFE standards. Like the tax credits for electric vehicles, the other tax credits also had potential long-run effects. The credits for traditional hybrids could have a lasting impact on CAFE standards, and the credits for biofuels could affect the commercial viability of those alternative fuels over the long term. Such potential effects make quantifying the overall cost-effectiveness of those other tax credits difficult. One way to compare the cost-effectiveness of the electric vehicle credits with that of the other programs is to evaluate it on the basis of the short-term effects that the credits would have if CAFE standards were not in force. That approach is equivalent to assuming that the reductions in gasoline use and greenhouse gas emissions resulting directly from electric vehicles purchased because of the credits equal the reductions resulting from induced changes in future CAFE standards. More generally, short-term effects are also relevant in comparing the credits for electric vehicles with the credits for traditional hybrids if the reductions in fuel use and emissions associated with sales of the two types of vehicles have comparable influence on future CAFE standards. In those terms, the cost of the tax credits per gallon of gasoline saved through the sale and use of electric vehicles tends to be higher than the average cost of the "Cash for Clunkers" program or the biofuel credits (see Figure 6). Compared with the credit for traditional hybrid vehicles, the current credits can cost more or less per gallon saved, depending on an electric vehicle's size and type. For example, they cost less when given for purchases of plug-in hybrids that are comparable in size to conventional vehicles with average vehicle fuel economy and have 4 kWh or 16 kWh batteries, but they cost more when given for all-electric vehicles of that size with 24 kWh batteries.[28]

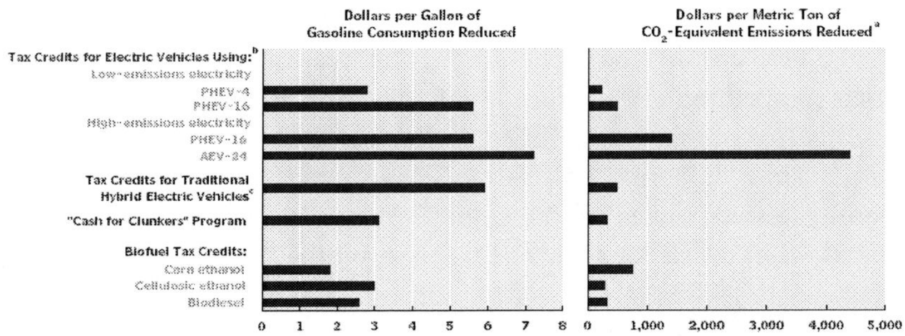

Source: Congressional Budget Office based on its own calculations (for the various tax credits) and on estimates from Shanjun Li, Joshua Linn, and Elisheba Beia Spiller, *Evaluating "Cash-for-Clunkers": Program Effects on Auto Sales and the Environment*, RFF Discussion Paper 10-39-REV (Resources for the Future, October 2011).

Notes: This figure compares what the short-term costs of various policies would be if the federal government's corporate average fuel economy standards were not in force, and if there was no indirect price-competition effect.

CO_2 = carbon dioxide; PHEV-4 = plug-in hybrid electric vehicle with a 4 kilowatt-hour (kWh) battery; AEV-24 = all-electric vehicle with a 24 kWh battery.

[a] Because individual greenhouse gases vary in their warming characteristics and persistence in the atmosphere, researchers commonly measure emissions in kilograms or metric tons of carbon dioxide equivalent—the amount of carbon dioxide that would cause an equivalent amount of warming over 100 years.

[b] These numbers apply to electric vehicles that substitute for average-fuel-economy conventional vehicles of similar size and performance. The costs of the electric vehicle tax credits can be higher or lower for other classes of vehicles (such as low-fuel-economy light-duty trucks or high-fuel-economy compact cars). Low-emissions electricity is electricity generated from nuclear power or renewable energy sources; high-emissions electricity is electricity generated from coal.

[c] CBO's analysis of the earlier tax credits for traditional hybrid vehicles uses the same assumptions as its analysis of the current tax credits for electric vehicles, with two exceptions: Estimates of the average tax credit provided on traditional hybrid vehicles and estimates of the share of vehicle sales attributable to those tax credits come from Arie Beresteanu and Shanjun Li, "Gasoline Prices, Government Support, and the Demand for Hybrid Vehicles in the United States," *International Economic Review,* vol. 52, no. 1 (February 2011).

Figure 6. Cost to the Federal Government of Using Various Policies to Reduce Gasoline Consumption and Greenhouse Gas Emissions in the Transportation Sector.

Taking account of the subsidy programs' indirect price-competition effects changes the comparisons somewhat. CBO estimates that the credits for electric vehicles will spur enough additional sales of other high-fuel-efficiency vehicles to add about 50 percent to the credits' direct effect, thus reducing their per-gallon cost by about one-third. Other researchers have concluded that the price-competition effect added about 150 percent to the direct effect of the tax credits for traditional hybrids, reducing their per-gallon cost by 60 percent.[29] Thus, in terms of relative cost-effectiveness, the indirect price-competition effect increases the appeal of the electric vehicle tax credits compared with "Cash for Clunkers" or the biofuel credits but reduces their appeal compared with the credits for traditional hybrids.

The cost of the electric vehicle tax credits per ton of emissions reduced depends greatly on the amount of carbon released from power plants. That cost tends to be comparable with, or lower than, the cost of the other subsidy programs when the electricity used to charge electric vehicles comes from low-carbon sources but is far higher than the cost of those programs when electricity comes from high-carbon sources (see Figure 6). With average-emissions electricity, the cost per ton of emissions reduced is generally somewhat higher than the cost of the other programs when only the direct effects are considered but is more comparable when the indirect price-competition effect is included.

POSSIBLE APPROACHES FOR FUTURE POLICIES

Lawmakers could take a number of different approaches to setting future policies aimed at reducing gasoline consumption or greenhouse gas emissions. Some approaches would involve modifying the existing tax credits, whereas others might involve very different policies.[30]

Because the total amount of gasoline consumption and greenhouse gas emissions in the transportation sector depends largely on CAFE standards, modifying the tax credits for electric vehicles would probably have little impact in the short run. Such changes could have longer-term effects if they influenced future CAFE standards. Even then, however, some other federal policies could probably reduce gasoline consumption or greenhouse gas emissions at a lower cost to the federal government than the tax credits for electric vehicles could.

Changing the Size of the Electric Vehicle Tax Credits

Increasing or decreasing the dollar amount of the tax credit provided on each sale of a new electric vehicle would produce little change in greenhouse gas emissions or gasoline consumption while current CAFE standards remain in place. Thus, in the short term, reducing or eliminating the tax credits would reduce budget deficits with little or no harm to the pursuit of energy or environmental goals.

The long-term implications of changing the tax credits are less clear, however. As in the short run, increasing or decreasing the size of the credits would have a direct impact on the budgetary cost of the credits. For example, reducing the size would lower the cost of each credit claimed; it would probably also reduce the number of electric vehicles purchased and hence the number of credits claimed. However, the long-term effects of such changes would depend on how they influenced future CAFE standards. Gasoline consumption and greenhouse gas emissions could increase (rather than remain unchanged, as they would in the short run) if a reduction in purchases of electric vehicles led policymakers to set future CAFE standards lower than they would otherwise. The effect on CAFE standards could be long-lasting if the lower vehicle sales adversely affected the development of the U.S. electric vehicle industry. Increasing the size of the credits would have the opposite effects. Because there is no basis for predicting how near-term sales of electric vehicles will affect future CAFE standards, CBO cannot judge whether changing the size of the credits would have a net positive or negative impact on their cost-effectiveness over the long term.

Changing the Number of Electric Vehicle Tax Credits Available

Currently, the tax credits are available in full on the first 200,000 electric vehicles sold by each manufacturer for use in the United States. Increasing the limit on the number of new-vehicle sales eligible for the credits would probably have little effect in the short run (even if the stringency of CAFE standards was not the main determinant of overall gasoline consumption and greenhouse gas emissions in the transportation sector during that period). The reason is that such an increase would have an impact on energy and environmental benefits or the government's costs only after a large number of electric vehicles had been sold. And with total U.S. sales of such vehicles numbering only about 40,000 so far, reaching a large sales volume would probably take a number of years. Similarly, a small or moderate reduction in the number of tax credits available would have little immediate effect, unless buyers interpreted the lesser support as increasing the risk that the electric vehicle industry would fail.

The impact over the longer term (after the sales milestones are met) of a change in the number of tax credits available is uncertain. Like the effects of the existing credits, that impact depends on how future CAFE standards are set. For example, if doubling the number of tax credits available doubled the effect of the credits in increasing future CAFE standards, then the long-term cost-effectiveness of the additional credits would be similar to that of the existing credits.

The relationship between the number of tax credits available and the future CAFE standards set by policymakers depends on the relationship between the number of credits and the evolution of the electric vehicle industry. On the one hand, the energy and environmental benefits of increasing the number of tax credits could have a much higher cost to the government (per unit of fuel consumption or emissions reduced) than the benefits of the existing tax credits do if a self-sustaining electric vehicle industry would have developed even without such an increase. In that case, additional tax credits would have a small effect—and their unit cost to the government would be correspondingly high—because many of the additional vehicle sales would have occurred anyway, and thus there would be little impact on the expectations of policymakers setting CAFE standards. On the other hand, if the increase in the number of credits eventually had a large enough effect on sales to promote a self-sustaining electric vehicle industry when one would not have developed otherwise, the additional credits would have a much lower unit cost to the government than the existing credits do.

Equalizing Purchase Incentives for All Buyers of Electric Vehicles

Because the tax credits reduce a buyer's federal income tax liability on a dollar-for-dollar basis up to the value of the credit or the buyer's tax liability, whichever is smaller, most taxpayers cannot take full advantage of the credits. For tax year 2011, CBO estimates that about 20 percent of potential tax filers (including people who do not file because they are below the income threshold) had federal income tax liability of at least $7,500 (the maximum tax credit), and only about 40 percent had liability of $2,500 or more (the minimum tax credit).

Federal incentives to purchase electric vehicles could be structured differently to provide the same dollar value to all vehicle purchasers, regardless of their federal income tax liability. One alternative would be to keep the credits but make them fully refundable. With a refundable credit, taxpayers whose liability is too small to allow them to claim the full value of the credit would receive the difference in the form of a refund.

Another alternative would be to replace the tax credits with a direct, point-of-purchase rebate. A direct rebate, like the one used in the "Cash for Clunkers" program, could be handled entirely by the seller and applied to the purchase price of the vehicle.[31] A point-of-sale rebate would have additional advantages for buyers, including those who have enough tax liability to claim the full value of the current credits. For example, it would allow buyers whose purchases involved loans to reduce the amount they financed.[32]

Whether increasing the value of the incentive to some taxpayers would significantly affect sales of electric vehicles is unclear. Although most taxpayers cannot take full advantage of the current nonrefundable credits, higher-income households account for the majority of sales of new vehicles—and presumably for an even larger share of sales of more-expensive vehicles.[33] Thus, most purchases of electric vehicles in coming years will probably be made by people who have enough tax liability to apply the full value of the tax credit.

To the extent that making the credits refundable or changing to a point-of-sale rebate resulted in higher sales of electric vehicles, federal costs would rise. However, if such changes resulted in long-term reductions in gasoline use and greenhouse gas emissions by influencing CAFE standards, those reductions would tend to be more cost-effective than the reductions achieved by increasing the size of the current credits. Larger credits would raise the government's cost for each electric vehicle that would have been purchased with the current credits (except those for which the value of the credit was constrained by the buyer's tax liability). Making the incentives more valuable to buyers with modest tax liability through a rebate or a refundable credit would be less likely to increase federal costs for vehicle purchases that would have occurred in any event; in particular, it would not raise the cost of sales to buyers who had enough tax liability to claim the full credit.

Another difference between point-of-purchase rebates and tax credits is that rebates would require appropriations from the Congress, which would subject the program to more frequent review. Prospective buyers of electric vehicles might view the rebate program as having less certain federal support and might accelerate their purchases, increasing near-term sales at the expense of medium- and long-term sales.

Other Policies to Reduce Gasoline Use or Greenhouse Gas Emissions

Giving people incentives to purchase electric vehicles is only one way to pursue the goals of lowering gasoline consumption and greenhouse gas emissions. Other possible approaches include increasing gasoline taxes and imposing limits on greenhouse gas emissions from multiple sectors of the economy.

Raising Gasoline Taxes

Increasing taxes on gasoline— which currently average 49 cents per gallon, counting federal, state, and local taxes—would have an immediate effect on fuel use and greenhouse gas emissions, because consumers would drive less in the vehicles they already own.[34] The reduction in miles driven, which would not have an offsetting CAFE effect, would grow over time, as workers adjusted their commuting patterns. Higher gasoline taxes would also shift consumers' demand for new vehicles from low-fuel-economy models to higher-fueleconomy models (including electric vehicles). However, that shift would probably be offset by the CAFE effect, with manufacturers and retailers reducing the prices of low-fuel-economy vehicles to maintain some lower level of sales.

Unlike the incentives mentioned above, higher gasoline taxes would boost federal revenues and thus cannot be analyzed in terms of the government's cost per unit of gasoline saved or greenhouse gas emissions reduced. (Gasoline taxes have other implications for transportation and the economy that are beyond the scope of this report.) However, it is possible to calculate the amount of the increase in gasoline taxes that would reduce gasoline use or emissions in the long run by as much as tax credits that had a specific impact on the prevalence of electric vehicles. For instance, a tax increase of between about 30 cents and 55 cents per gallon (equivalent to a tax on carbon dioxide of $35 to $60 per metric ton) would reduce gasoline consumption by average-fuel-economy conventional vehicles to about the same extent that the tax credits would if they raised electric vehicles' share of total light-duty vehicles driven by 5 percentage points— an increase greater than the total share of traditional hybrids today.[35] The equivalent tax increase for reducing greenhouse gas emissions is somewhat smaller—in the range of 20 cents to 30 cents per gallon—because electric vehicles have a proportionately greater impact on gasoline consumption than they do on emissions. Those equivalent tax increases would be higher if electric vehicles were to account for a larger share of vehicles driven in the future.

Policies That Reduce Greenhouse Gas Emissions in Multiple Sectors

For the goal of reducing emissions, policies that apply to multiple sectors of the economy can be much more efficient than approaches that focus only on the transportation sector. One such option is a capand-trade program, in which policymakers set annual limits (caps) on total emissions; require entities regulated by the program (such as power plants and other large sources of emissions) to hold rights, or allowances, to emit greenhouse gases; and allow the entities to buy and sell allowances (the trade part of the program). Such a program would tend to minimize the total cost of achieving a given reduction in emissions, because regulated entities that would face relatively high costs to decrease their emissions could purchase allowances from other regulated entities that could reduce emissions more cheaply.[36]

Another broad option is to levy a tax on the carbon content of fossil fuels, with the size of the tax dependent on the amount of carbon dioxide released from burning those fuels. Fossil fuels are used in all sectors of the economy, either directly or indirectly (as an input to electricity), so a carbon tax would reduce emissions throughout the economy. Such reductions would come partly from increases in the use of fossil fuels with lower carbon intensity (such as natural gas) in place of fuels with higher carbon intensity (such as coal) and partly from decreases in the overall use of fossil fuels.

The advantage of a multisector approach can be seen by comparing the transportation and electricity sectors. The transportation sector is likely to be one of the most expensive in which to reduce emissions, because it is highly dependent on petroleum and few substitutes are available. The electricity sector, by contrast, has a much more diverse set of energy sources. Of the energy used to produce electricity in the United States, about 50 percent comes from coal, about 20 percent from natural gas, about 20 percent from nuclear power, and about 10 percent from renewable sources. The diversity of energy sources and the relative ease of substituting production from some power plants for production from others allow greater flexibility to use substitute fuels and to reduce emissions at lower cost.

For example, companies that produce energy from renewable sources are eligible for a tax credit of 2.1 cents per kWh produced. The cost to the federal government of that credit is about $8 per metric ton of CO_2e emissions reduced for energy produced from geothermal sources and about $12 per metric ton of CO_2e emissions reduced for energy from wind, compared with costs in the hundreds of dollars per metric ton for the transportation-related tax credits.[37] However, compared with emissions reductions in the transportation sector, reductions in the electricity sector have fewer ancillary benefits in terms of improving energy security by decreasing petroleum use.[38] They would also have less impact than the tax credits on the development of the electric vehicle industry.

APPENDIX: DETAILS OF THE TECHNICAL ASSUMPTIONS OF CBO'S ANALYSIS

The extent to which the cost to consumers of owning and operating an electric vehicle differs from that of a conventional vehicle or traditional hybrid depends on the vehicle's purchase price, its fuel economy, fuel prices, the weights that consumers apply to costs that occur at different times, and other factors. Vehicles with larger batteries are more expensive to buy, but they can be driven farther on electric power. The savings in fuel costs from using an electric vehicle depend on how much it is driven, the percentage of miles that it operates on electric power, its fuel efficiency relative to that of alternative vehicles, and the prices of electricity and gasoline (or other liquid fuels). The estimates and assumptions about those factors that the Congressional Budget Office (CBO) used in this analysis are specified below.

Prices for Electric Vehicles

Because the tax credits for the purchase of an electric vehicle and the price that consumers pay for that vehicle depend on the size of the battery, CBO estimated the price difference between electric and conventional vehicles on the basis of the electric vehicle's battery capacity, measured in kilowatt-hours (kWh) of electricity, along with other factors that affect that difference.

On the basis of numerous studies about the current and projected prices of electric vehicles, CBO concluded that the average difference in retail price between a plug-in hybrid vehicle and an equivalent conventional vehicle includes a fixed component of about $4,000, which represents cost differences that are independent of battery size, and a variable component of about $950 per kWh of battery capacity, which represents costs that vary with the size of the vehicle's battery. (Those figures include a 40 percent markup to reflect the difference between the cost of producing those vehicles and the price that consumers pay for them.)[1] At those prices, a 16 kWh plug-in hybrid costs about $19,000 more than a comparable conventional vehicle.[2]

Because different classes of electric vehicles cost more or less to produce, CBO estimates that the average additional costs (both fixed and variable) of producing an electric version of a light-duty truck are about 10 percent higher than the average additional costs of all vehicles, whereas the additional costs of producing an electric passenger car are about 10 percent lower than the overall average. (Light-duty trucks include pickup trucks, minivans, and sport-utility vehicles with a gross weight of no more than 8,500 pounds—the definition of light duty.)

The cost differential between electric and conventional vehicles is about 40 percent lower for all-electric vehicles—which are assumed to have 24 kWh of battery capacity—than for plug-in hybrid vehicles with the same battery size, in part because all-electric vehicles use no gasoline and therefore do not require an internal combustion engine or an exhaust system.[3] In addition, all-electric vehicles have simpler transmission and fueling systems. However, all-electric vehicles generally require larger batteries to ensure that they can be driven far enough to satisfy consumers' needs. (Unlike a plug-in hybrid vehicle, which can use gasoline like a conventional vehicle once its battery power is exhausted, an all-electric vehicle must stop and be recharged, a process that takes far longer than filling a gasoline tank.) The need for large batteries can be significant enough that some all-electric vehicles will be more costly than smaller plug-in hybrid vehicles, despite their lower average cost per unit of battery capacity.

Vehicles' Fuel Economy

To ensure consistency in estimates of fuel economy for electric vehicles and for comparable conventional and traditional hybrid vehicles, CBO relied on estimates for three different hypothetical classes of electric vehicles: low-fuel-economy light-duty trucks, average-fueleconomy light-duty vehicles, and high-fuel-economy compact cars.[4] The Electric Power Research Institute

(EPRI) estimates that those classes of vehicles could travel between 2 and 4 miles per kWh of electricity when running on electric power (including electricity lost when recharging the plug-in batteries) and, in the case of plug-in hybrids, could travel between 20 and 60 miles per gallon when running on gasoline. Electric versions of low-fuel-economy light-duty trucks would be at the lower end of those ranges, and high-fuel-economy compact cars would be at the upper end.

Traditional hybrid vehicles and plug-in hybrids have about 50 percent higher fuel economy when running on gasoline than equivalent conventional vehicles do.[5] That difference occurs because hybrids use their more efficient electric motor during acceleration and because the energy released during braking is recycled to recharge the battery. Because that greater fuel efficiency is a product of the number of starts and stops that a vehicle makes, the effect is larger for city travel than for highway travel, a distinction not considered in this analysis.

Depending on vehicle class, the fuel economies of conventional vehicles considered in this analysis initially range from 15 miles per gallon to about 40 miles per gallon, with average conventional vehicles having a fuel efficiency of about 25 miles per gallon—a value close to the current average fuel economy standard for automakers as a group. However, the corporate average fuel economy standards for all light-duty vehicles that a manufacturer sells in a given year are set to increase by an average of 3.5 percent a year from 2012 to 2021.[6] In evaluating the future cost-effectiveness of the tax credits, CBO assumed that the fuel economies of all new conventional vehicles, traditional hybrids, and plug-in hybrids sold would increase at that rate through 2020.

Fuel Prices

CBO assumed that retail prices for gasoline would rise in real terms (that is, adjusted to remove the effects of inflation) over time but that electricity prices would remain essentially unchanged in real terms. For simplicity, CBO assigned average real prices of gasoline and electricity over the life of a given vehicle. Thus, for example, the gasoline used by current-model vehicles is assumed to have a constant real price of about $3.60 per gallon, whereas the gasoline used by new 2020 vehicles is assumed to have a real price of about $3.90 per gallon. For electricity, the real price changes only slightly and thus remains at around 12 cents per kWh for vehicles in both model years. Those figures, which are in 2010 dollars, are based on recent projections by the Energy Information Administration.[7]

Discount Rate

Because money received in the future is valued less than the same amount received today, CBO discounted fuel costs and tax credits to reflect the time at which they are projected to be paid or received. Those discounted amounts allowed CBO to estimate the present value of the total lifetime costs of purchasing and operating a particular type of vehicle.[8]

To discount savings from lower gasoline use, CBO used a rate of 10 percent per year, which is consistent with evidence on how consumers discount future savings on energy when making decisions about other types of purchases, such as home appliances or energy-efficiency upgrades.[9] That rate is higher than the discount rates that CBO has used for future fuel purchases in other contexts.[10] One reason for such a relatively high rate is that there is considerable uncertainty about how higher vehicle fuel economy will reduce fuel costs in the future. The extent of that reduction can depend on people's

individual circumstances—such as how long they will own a vehicle, how extensively they will drive it, and their local weather and driving conditions—as well as broader factors such as how reliable electric vehicles and their batteries will be and how high fuel prices will rise in the future. Another reason for the relatively high discount rate may be that consumers tend to focus on the attributes of a vehicle that are most salient at the time of purchase, such as the purchase price, size, comfort, safety, and performance. Because those factors may matter differently to different buyers, CBO also considered the impact of discount rates that were half as high (5 percent) and twice as high (20 percent) on the results of its analysis.

Discounting also slightly reduces the value to consumers of the tax credits. Because the receipt of a tax credit is much more certain to a new-vehicle buyer than future reductions in gasoline costs are, CBO discounted the tax credit at 2 percent per year, a rate consistent with expected short-term interest rates in coming years. CBO assumed that the credit would be applied about eight months after an electric vehicle was purchased, on average, so the 2 percent discount rate implied roughly a 1 percent reduction in the perceived value of the credit.

Other Technical Assumptions

In this analysis, all types of vehicles (except all-electric vehicles, as discussed below) are assumed to have an operating life of 150,000 miles. In addition, maintenance and repair costs are assumed to be the same for all vehicles, because CBO had no basis to judge whether electric vehicles would be more or less reliable or costly to repair than conventional vehicles. Thus, fuel costs are the only factor in the analysis that creates differences in operating expenses.

The average percentage of miles that a plug-in hybrid vehicle operates on electric power depends on the number of miles it is driven each day and the maximum distance it can travel on a single charge without using gasoline (the vehicle's electric range). On the basis of an analysis of the reported driving habits of U.S. drivers, CBO assumes that the plug-in hybrid vehicles in its analysis can operate on electric power for 20 percent to 70 percent of the miles they travel, given the range of battery capacities considered in this study.[11] Those percentages include all of the miles traveled by people who drive less than their vehicle's electric range in a given day and all of the miles not exceeding the vehicle's electric range traveled by people who drive longer distances. If, for example, 10 drivers each drove 10 miles and 5 drivers each drove 30 miles, electric power could account for 70 percent of the 250 miles traveled by electric vehicles with a 15-mile electric range (all 100 miles of travel by people driving 10 miles and 75 miles of travel by people driving 30 miles). Increasing either the frequency with which plug-in hybrids were recharged or the size of their batteries would raise that percentage.

With all-electric vehicles, electric power accounts for 100 percent of travel, but mileage is constrained by the lack of a secondary fuel source. CBO estimates the distance that all-electric vehicles travel each year as fractions of the distance traveled by other comparable vehicles (conventional vehicles, traditional hybrids, and plug-in hybrids). Those fractions are derived by comparing the average daily travel for vehicles that are not driven extensively with the average for all vehicles. For example, the maximum distance (or range) that an average all-electric vehicle in CBO's analysis can drive without recharging is about 55 miles, and the average distance traveled by drivers each day whose travel does not exceed 55 miles is about 21 miles—or about 45 percent lower than the 38-mile national average for all light-duty vehicles.[12] The corresponding averages for all-electric versions of light-duty trucks and compact cars are roughly 16 miles and 24 miles driven per day, respectively, or about 60 percent and 35 percent lower than the average for all light-duty vehicles.

In assessing the lifetime costs of electric vehicles and the emissions associated with their use, CBO took into account the amount of electricity lost between its generation and its use for travel. An estimated 7 percent of electric power is lost in the transmission and distribution system—that is, between generating facilities and consumers.[13] An additional 12 percent is lost when charging an electric vehicle and converting the alternating current produced by power plants into the direct current stored in the vehicle's battery.[14]

How Tax Credits Affect Sales of Electric Vehicles and Overall Fuel Economy

As of yet, no reliable estimates exist of the share of electric vehicle sales that can be attributed to the tax credits— that is, of the number of vehicles that would not have been sold without those credits. However, researchers have analyzed the earlier tax credits that applied to purchases of traditional hybrids (which expired at the end of 2010) and estimated that they were responsible for roughly one-quarter of all sales of those vehicles.[15] On the basis of that research, CBO estimates that the current tax credits will be responsible for 30 percent of all electric vehicles sold. That share is slightly higher than the estimate for the earlier tax credits because the current credits are larger (a maximum of $7,500 rather than $3,400) relative to the cost of the vehicles.

The effects of the tax credits may extend beyond additional sales of electric vehicles, however. By lowering the effective purchase price of electric vehicles, the credits are likely to cause manufacturers to reduce prices of competing high-fuel-economy conventional and traditional hybrid vehicles, thus raising the average fuel economy of other, nonelectric vehicles sold. That effect may be large. The results of one study on the energy and environmental gains from the tax credits for traditional hybrids suggest that the benefits from increased sales of higher-fuel-economy conventional vehicles were about one and a half times the benefits from sales of the hybrids themselves.[16]

The relative size of those various gains depends on how many vehicles are sold. Low sales of electric vehicles would not greatly affect the average fuel economy of conventional vehicles. At least in the near term, sales of electric vehicles will probably not be as large as sales of traditional hybrids have been—because of differences in their costs—so the impact of electric vehicles on the average fuel economy of other vehicles will probably be smaller than was found for traditional hybrids. CBO assumes that those spillover effects will be about half as large as the direct effect from sales of electric vehicles. Thus, in CBO's analysis, the total reductions in fuel use and greenhouse emissions from electric vehicles are 1.5 times the direct reductions.

End Notes for Summary

[1] Light-duty vehicles include passenger cars and light-duty trucks (such as pickup trucks, minivans, and sport-utility vehicles) that have a gross weight of no more than 8,500 pounds.

[2] Because individual greenhouse gases vary in their warming characteristics and persistence in the atmosphere, researchers commonly measure emissions in kilograms or metric tons of carbon dioxide equivalent—the amount of carbon dioxide that would cause an equivalent amount of warming over 100 years.

[3] The National Highway Traffic Safety Administration (NHTSA) sets CAFE standards for the fuel economy of passenger cars and light-duty trucks, and the Environmental Protection Agency (EPA) sets parallel standards to limit emissions of greenhouse gases from such vehicles. NHTSA is allowed to set CAFE standards for only five years at a time (currently, through model year 2021); EPA has set emissions standards that run through model year 2025.

[4] Of the tax credits for biofuels, those for corn ethanol and biodiesel expired in 2011, and the credits for cellulosic ethanol are due to expire at the end of 2012.

[5] For example, two bills introduced in the current (112th) Congress (S. 232 and H.R. 500) would make the credits apply to the first 500,000 electric vehicles sold by each manufacturer (up from 200,000 under current law).

End Notes

[1] For more information about those standards, see Box 2 on page 14 and Congressional Budget Office, *How Would Proposed Fuel Economy Standards Affect the Highway Trust Fund?* (May 2012).

[2] State tax credits for the purchase of new electric vehicles can be as high as $7,500, although only a handful of states offer tax credits.

[3] Joint Committee on Taxation, Estimated Budget Effects of the Revenue Provisions Contained in the Conference Agreement for H.R. 1, the "American Recovery and Reinvestment Tax Act of 2009," JCX-19-09 (February 12, 2009), p. 3.

[4] Department of Energy, "Recovery Act Awards for Electric Drive Vehicle Battery and Component Manufacturing Initiative" (October 2011), www1.eere.energy battery_ awardee_list.pdf.

[5] Department of Energy, *One Million Electric Vehicles by 2015: February 2011 Status Report* (February 2011), www1.eere .energy_ vehicles_rpt.pdf.

[6] Department of Energy, "Recovery Act Awards for Electric Drive Vehicle Battery and Component Manufacturing Initiative" (October 2011), www1.eere.energy _awardee_list.pdf.

7 For a list of those loans, see the Web site of the Department of Energy's Loan Programs Office, http://lpo.energy.gov/ ?page_id=45. The $8.4 billion total differs from what was reported for the ATVM program in Congressional Budget Office, *Federal Financial Support for the Development and Production of Fuels and Energy Technologies* (March 2012), because a $730 million loan that had previously been conditionally approved by DOE was not granted.

8 The timing of future ATVM loans is unknown, so some of the estimated $1.5 billion in future subsidy costs could be incurred after 2019.

9 Light-duty vehicles include passenger cars and light-duty trucks (such as pickup trucks, minivans, and sport-utility vehicles) that have a gross weight of no more than 8,500 pounds. Most of the electric vehicles that are currently available are midsize or compact passenger cars. Although no plug-in hybrid or all-electric light-duty trucks are on the market now, such vehicles may become available in the next few years.

10 A present value is a single number that expresses a flow of current and future costs or savings in terms of an equivalent lump sum paid or saved today. The present value depends on the rate of interest (the discount rate) that is used to translate future cash flows into current dollars.

11 Because all-electric vehicles have no secondary fuel source and cannot travel long distances before being recharged, CBO assumes that they will be driven roughly half as far each year as other vehicles. If everything else is equal, that reduced driving range increases the relative operating costs of all-electric vehicles, because fewer miles traveled means smaller savings in fuel costs.

12 Of the notional vehicles included in CBO's analysis, only a plug-in hybrid light-duty truck with a 4 kWh battery could be cost-competitive with a comparable conventional vehicle today even without the tax credit. However, plug-in hybrid light-duty trucks with such small batteries are unlikely to be produced for sale; in general, those trucks will probably require larger-capacity batteries, given their size and performance. CBO included such vehicles in its analysis to help show how vehicle type and battery size affect the results.

13 At some point, continued declines in the prices of electric vehicles, combined with their lower fuel costs, could allow them to be cost-competitive with conventional vehicles even without the tax credits. On the basis of the experts' projections, CBO projects that most electric vehicles will not reach that point until after 2030, though developments that far in the future are very uncertain.

14 Those estimates of the additional fixed and variable costs of electric vehicles are averages for all light-duty vehicles. In CBO's analysis, those additional costs are about 10 percent higher than average for light-duty trucks and about 10 percent lower than average for passenger cars (see the appendix).

15 The approximately $16,000 additional cost of an all-electric vehicle (the purchase price of that vehicle minus the purchase price of an equivalent conventional vehicle) is about 60 percent of the approximately $27,000 additional cost of a 24 kWh plug-in hybrid [$4,000 + (24 x $950)].

16 Energy Information Administration, *Annual Energy Outlook 2012,* DOE/EIA-0383(2012) (June 2012). CBO used a higher gasoline price for its analysis of vehicles sold in 2020 (see the appendix).

17 See, for example, Mark K. Dreyfus and W. Kip Viscusi, "Rates of Time Preference and Consumer Valuations of Automobile Safety and Fuel Efficiency," *Journal of Law and Economics,* vol. 38, no. 1 (April 1995), pp. 79–105; Hunt Allcott and Nathan Wozny, *Gasoline Prices, Fuel Economy, and the Energy Paradox,* 10-003 (Massachusetts Institute

of Technology, Center for Energy and Environmental Policy Research, March 2010); and Thomas S. Turrentine and Kenneth S. Kurani, "Car Buyers and Fuel Economy?," *Energy Policy*, vol. 35, no. 2 (February 2007), pp. 1213–1223.

18 There is some evidence that people discount future savings from improvements in energy efficiency at rates higher than 10 percent—high enough, for example, that they effectively take into account only three years' worth of fuel costs when deciding whether to buy a vehicle with high fuel economy. See David L. Greene, John German, and Mark A. Delucchi, "Fuel Economy: The Case for Market Failure," in Daniel Sperling and James S. Cannon, eds., *Reducing Climate Impacts in the Transportation Sector* (Springer, 2009). Such high discount rates could reflect consumers' belief that long-term savings cannot be relied on because factors such as future gasoline prices and the maintenance and repair costs of electric vehicles are too uncertain.

19 Because individual greenhouse gases vary in their warming characteristics and persistence in the atmosphere, researchers commonly measure emissions in kilograms or metric tons of carbon dioxide equivalent—the amount of carbon dioxide that would cause an equivalent amount of warming over 100 years.

20 For estimates of the benefits to society from reducing gasoline consumption, see Congressional Budget Office, *Alternative Approaches to Funding Highways* (March 2011).

21 Other factors that could affect the cost to the government of the electric vehicle tax credits include changes in receipts from gasoline taxes and changes in costs for road repairs because of changes in driving behavior. Those factors are beyond the scope of this analysis.

22 Arie Beresteanu and Shanjun Li, "Gasoline Prices, Government Support, and the Demand for Hybrid Vehicles in the United States," *International Economic Review,* vol. 52, no. 1 (February 2011), pp. 161–182.

23 See Virginia McConnell and Tom Turrentine, *Should Hybrid Vehicles Be Subsidized?* Backgrounder (Resources for the Future and National Energy Policy Institute, July 2010); and W. Ross Morrow and others, "Analysis of Policies to Reduce Oil Consumption and Greenhouse-Gas Emissions from the U.S. Transportation Sector," *Energy Policy*, vol. 38, no. 3 (March 2010), pp. 1305–1320.

24 That net impact might not be zero, however, for several reasons. Two involve details about how CAFE standards are implemented. First, the National Highway Transportation Safety Administration and the Environmental Protection Agency (EPA) model all-electric vehicles as having a high miles-per-gallon rating, calculated using a "petroleum equivalency factor," even though those vehicles do not use any liquid fuel. (The same treatment applies to the electricity used by plug-in hybrids.) Treating such vehicles as though they use some petroleum rather than none reduces their impact on the average fuel efficiency of automakers' fleets, as calculated for the purposes of CAFE compliance, thereby limiting companies' ability to increase their sales of low-fuelefficiency vehicles and still meet the standards. Second, for purposes of EPA's standards for greenhouse gas emissions, electric vehicles are weighted more than conventional vehicles in calculating the average emissions of a manufacturer's fleet. The practical effect of that weighting is limited by the fact that it is not used in evaluating compliance with CAFE standards. Another reason that the net impact of the credits on energy use and emissions might not be zero is that total miles driven may change with changes in the mix of vehicles purchased, even if overall fuel efficiency remains constant, because drivers' responses to their vehicles' higher or lower fuel costs may not offset each other.

25 NHTSA did not take the effect of plug-in hybrids into account for model years before 2020. A provision of the law under which the agency sets fuel economy standards for light-duty

vehicles requires it to consider dual-fueled automobiles to be operated only on gasoline or diesel fuel. [See 49 U.S.C. §32902(h)(2) (2006)]. According to NHTSA, the purpose of the law governing dual-fueled automobiles is better served by interpreting that provision as moot for model years after 2019, because current statutory flexibilities relating to dual-fueled automobiles expire after that year. See *2017 and Later Model Year Light-Duty Vehicle Greenhouse Gas Emissions and Corporate Average Fuel Economy Standards,* 76 Fed. Reg. 75,854, 75,226 (Dec. 1, 2011).

[26] See *Light-Duty Vehicle Greenhouse Gas Emission Standards and Corporate Average Fuel Economy Standards,* 75 Fed. Reg. 25,324 (May 7, 2010); and Environmental Protection Agency, National Highway Traffic Safety Administration, and California Air Resources Board, *Interim Joint Technical Assessment Report: Light-Duty Vehicle Greenhouse Gas Emission Standards and Corporate Average Fuel Economy Standards for Model Years 2017–2025* (September 2010), www.epa.gov/oms/climate ldv-ghg-tar.pdf.

[27] Biofuel tax credits do not affect the mix of new vehicles sold, so CAFE standards have no impact on the extent to which the use of biofuels reduces the use of petroleum-based fuels. With the "Cash for Clunkers" program, although the purchase of a new higher fuel-economy vehicle might have allowed automakers to sell an additional low-fuel-economy vehicle under the CAFE standards that were in effect at the time, overall fuel use would decline because the fuel efficiency of the "clunker" taken out of service was probably lower than the average required for all new vehicles.

[28] In calculating the cost to the government of the tax credits formerly available for traditional hybrid vehicles, CBO used the same assumptions that governed its evaluation of the electric vehicle tax credits, with two exceptions: The average tax credit was estimated to be about $2,500 (in 2010 dollars, based on the average tax credit provided for traditional hybrid sales in 2006), and the tax credit was estimated to be responsible for 20 percent of traditional hybrids sold (the corresponding figure for electric vehicles was 30 percent). See Arie Beresteanu and Shanjun Li, "Gasoline Prices, Government Support, and the Demand for Hybrid Vehicles in the United States," *International Economic Review,* vol. 52, no. 1 (February 2011), pp. 161–182.

[29] See Arie Beresteanu and Shanjun Li, "Gasoline Prices, Government Support, and the Demand for Hybrid Vehicles in the United States," *International Economic Review,* vol. 52, no. 1 (February 2011), pp. 161–182.

[30] Bills introduced in the current (112th) Congress (such as S. 232 and H.R. 500) would increase the number of electric vehicle sales to which the tax credits would apply, effectively extending the credits. In addition, the President's budget for 2013 contained several proposals to alter the credits, including making them available for more vehicle sales and raising the maximum dollar amount from $7,500 to $10,000. See Department of the Treasury, *General Explanations of the Administration's Fiscal Year 2013 Revenue Proposals* (February 2012), pp. 32–33, www.treasury.gov/resourcecenter/tax-policy

[31] The President's budget for 2013 included a proposal to switch eligibility for the electric vehicle tax credits from buyers to sellers, allowing the sellers to offer point-of-sale rebates. Sellers would be required to disclose the amount of the credit available for each vehicle sold. See Department of the Treasury, *General Explanations of the Administration's Fiscal Year 2013 Revenue Proposals* (February 2012), pp. 32–33, www.treasury.gov/resource-center/tax-policy Pages/general_explanation.aspx.

[32] Another benefit of an immediate rebate is that it would have a greater discounted value to buyers than a credit that they did not receive until later, when they filed their tax returns. That difference in value would be very small, however (see the appendix).

[33] Laura Paszkiewicz, "The Cost and Demographics of Vehicle Acquisition," *Consumer Expenditure Survey Anthology, 2003,* Report 967 (Department of Labor, Bureau of Labor Statistics, September 2003), pp. 61–66.

[34] The American Petroleum Institute publishes estimates of average fuel taxes each calendar quarter; for the latest figures, see American Petroleum Institute, "Oil and Natural Gas Overview: Motor Fuel Taxes," www.api.org/oil-

[35] The estimates for the equivalent tax on gasoline are based on a long-term price elasticity of demand for gasoline of -0.4, meaning that a 10 percent increase in gasoline prices reduces gasoline consumption by 4 percent in the long run. (The demand for gasoline is less sensitive to price changes in the short run.) A plug-in hybrid vehicle with a 16 kWh battery, for example, would consume about 70 percent less gasoline than a conventional vehicle of comparable size and performance, so overall gasoline consumption would decline by roughly 5 percent if the market share for those vehicles increased by 5 percentage points (a 70 percent reduction for each plug-in hybrid driven, times the assumed 5 percentage-point increase in the share of all vehicles driven, times 1.5 for the indirect effect that sales of electric vehicles have on the average fuel economy of conventional vehicles sold). If achieved through an increase in gasoline taxes, that decline of roughly 5 percent in long-term gasoline consumption would require a 13 percent increase in gasoline prices (5 percent divided by the long-term price elasticity of demand of 0.4), which amounts to about 50 cents per gallon under CBO's assumption that the long-term average price of gasoline is $3.60 per gallon (in 2010 dollars). The range of 30 cents to 55 cents discussed above reflects the different sizes of batteries and types of electric vehicles evaluated.

[36] Estimates of the government's cost of using the tax credits for electric vehicles to reduce greenhouse gas emissions are not comparable with estimates of the price of allowances under a cap-and-trade program for reducing emissions. The cost reported in this study represents the average tax revenue forgone because of the tax credits, whereas the price of allowances under a capand-trade program represents the cost of avoiding the last (or marginal) ton of emissions under the specific requirements of the program. For more information about cap-and-trade programs for emissions, see Congressional Budget Office, *Managing Allowance Prices in a Cap-and-Trade Program* (November 2010).

[37] Gilbert E. Metcalf, "Tax Policies for Low-Carbon Technologies," *National Tax Journal,* vol. 62, no. 3 (September 2009), pp. 519– 533. Those costs to the government of the production tax credit for renewable energy are not directly comparable with the estimates in this study because they do not include an adjustment for the portion of the wind or geothermal production that would have occurred in the absence of the tax credit. Including such an adjustment—so that the government's costs reflected only the emissions reductions caused by the tax credit—would raise the costs of that credit. However, they would probably still be well below CBO's estimates for the electric vehicle tax credits.

[38] For a discussion of energy security, see Congressional Budget Office, *Energy Security in the United States* (May 2012).

End Notes for Appendix

[1] Lynette Cheah and John Heywood, "The Cost of Vehicle Electrification: A Literature Review" (paper presented at the Massachusetts Institute of Technology's Energy Initiative Symposium "The Electrification of the Transportation System: Issues and Opportunities,"

Cambridge, Mass., April 8, 2010); National Research Council, Committee on Assessment of Resource Needs for Fuel Cell and Hydrogen Technologies, *Transitions to Alternative Transportation Technologies—Plug-in Hybrid Electric Vehicles* (National Academies Press, 2010); and Boston Consulting Group, *Batteries for Electric Cars: Challenges, Opportunities, and the Outlook to 2020* (BCG, 2010), www.bcg.com/documents/file36615.pdf.

[2] Taking into account both the fixed and variable components of cost, the $19,000 corresponds to roughly $1,200 per kWh. That figure is more comparable with other analysts' estimates that do not distinguish between fixed and variable components.

[3] Anup Bandivadekar and others, *On the Road in 2035: Reducing Transportation's Petroleum Consumption and GHG Emissions,* LFEE 2008-05 RP (Massachusetts Institute of Technology, Laboratory for Energy and the Environment, July 2008); Steve Plotkin and others, *Multi-Path Transportation Futures Study: Vehicle Characterization and Scenario Analyses,* ANL/ESD/09-5 (Argonne National Laboratory, July 2009); Matthew A. Kromer and John B. Heywood, *Electric Powertrains: Opportunities and Challenges in the U.S. Light-Duty Vehicle Fleet,* LFEE 2007-03 RP (Massachusetts Institute of Technology, Laboratory for Energy and the Environment, May 2007); and Jeremy J. Michalek and others, "Valuation of Plug-In Vehicle Life-Cycle Air Emissions and Oil Displacement Benefits," *Proceedings of the National Academy of Sciences,* vol. 108, no. 40 (October 4, 2011), pp. 16554–16558.

[4] Electric Power Research Institute, Environmental Assessment of Plug-In Hybrid Electric Vehicles, vol. 2, United States Air Quality Analysis Based on AEO-2006 Assumptions for 2030, Technical Report 1015326 (EPRI, July 2007); and Electric Power Research Institute, Comparing the Benefits and Impacts of Hybrid Electric Vehicle Options for Compact Sedan and Sport Utility Vehicles, Technical Report 1006892 (EPRI, July 2002).

[5] Electric Power Research Institute, Environmental Assessment of Plug-In Hybrid Electric Vehicles, vol. 2, United States Air Quality Analysis Based on AEO-2006 Assumptions for 2030, Technical Report 1015326 (EPRI, July 2007); and Constantine Samaras and Kyle Meisterling, "Life Cycle Assessment of Greenhouse Gas Emissions from Plug-in Hybrid Vehicles: Implications for Policy," Environmental Science and Technology, vol. 42, no. 9 (May 1, 2008), pp. 3170–3176.

[6] Light-Duty Vehicle Greenhouse Gas Emission Standards and Corporate Average Fuel Economy Standards, 75 Fed. Reg. 25,324 (May7, 2010); and 2017 and Later Model Year Light-Duty Vehicle Greenhouse Gas Emission Standard and Corporate Average Fuel Economy Standards, __ Fed. Reg. __ (Aug. 28, 2012) (to be codified at 40 C.F.R. pts. 85, 86, and 600; 49 C.F.R. pts. 523, 531, 533, 536, and 537) (Final Rule to be published in the Federal Register).

[7] Energy Information Administration, *Annual Energy Outlook 2012,* DOE/EIA-0383(2012) (June 2012).

[8] CBO did not use discounting when it analyzed the cost-effectiveness of the tax credits in reducing gasoline consumption and greenhouse gas emissions, even though reductions that occur earlier may be somewhat more valuable (because the benefits begin sooner) than reductions that occur later. Although discounting is important in assessing people's choices about vehicle purchases, it would not affect the conclusions about the cost-effectiveness of alternative policies—in part because the effects of other policies that also work by changing the composition of the overall vehicle fleet are similarly spread out over time.

[9] See, for example, Mark K. Dreyfus and W. Kip Viscusi, "Rates of Time Preference and Consumer Valuations of Automobile Safety and Fuel Efficiency," *Journal of Law and*

Economics, vol. 38, no. 1 (April 1995), pp. 79–105; Hunt Allcott and Nathan Wozny, *Gasoline Prices, Fuel Economy, and the Energy Paradox,* 10-003 (Massachusetts Institute of Technology, Center for Energy and Environmental Policy Research, March 2010); and Thomas S. Turrentine and Kenneth S. Kurani, "Car Buyers and Fuel Economy?," *Energy Policy*, vol. 35, no. 2 (February 2007), pp. 1213–1223.

[10] For example, in a May 2011 study, *The Cost-Effectiveness of Nuclear Power for Navy Surface Ships*, CBO discounted future spending on oil supplies at 3 percent, a rate consistent with the estimated return that private-sector investors would require to provide fuel.

[11] SAE International, Utility Factor Definitions for Plug-In Hybrid Electric Vehicles Using 2001 U.S. DOT National Household Travel Survey Data, J2841 (SAE International, March 2009).

[12] Based on data from the Federal Highway Administration's 2009 National Household Travel Survey.

[13] Energy Information Administration, *Annual Energy Review 2010,* DOE/EIA-0384(2010) (October 2011), Figure 8.0, footnote 5.

[14] Electric Power Research Institute, Environmental Assessment of Plug-In Hybrid Electric Vehicles, vol. 2, United States Air Quality Analysis Based on AEO-2006 Assumptions for 2030, Technical Report 1015326 (EPRI, July 2007).

[15] Arie Beresteanu and Shanjun Li, "Gasoline Prices, Government Support, and the Demand for Hybrid Vehicles in the United States," *International Economic Review,* vol. 52, no. 1 (February 2011), pp. 161–182; Kelly Sims Gallagher and Erich Muehlegger, "Giving Green to Get Green? Incentives and Consumer Adoption of Hybrid Vehicle Technology," *Journal of Environmental Economics and Management,* vol. 61, no. 1 (January 2011), pp. 1–15; and Ambarish Chandra, Sumeet Gulati, and Milind Kandlikar, "Green Drivers or Free Riders? An Analysis of Tax Rebates for Hybrid Vehicles," *Journal of Environmental Economics and Management,* vol. 60, no 2 (September 2010), pp. 78–93.

[16] Arie Beresteanu and Shanjun Li, "Gasoline Prices, Government Support, and the Demand for Hybrid Vehicles in the United States," *International Economic Review,* vol. 52, no. 1 (February 2011), pp. 161–182.

In: Electric Vehicles and the Wind Industry ISBN: 978-1-62417-298-4
Editor: Adam P. Reyes

Chapter 2

U.S. RENEWABLE ELECTRICITY: HOW DOES THE PRODUCTION TAX CREDIT (PTC) IMPACT WIND MARKETS?*

Phillip Brown

SUMMARY

U.S. wind projects that use large turbines—greater than 100 kilowatts (kW)—are eligible to receive federal tax incentives in the form of production tax credits (PTC) and accelerated depreciation. Originally established in 1992, the PTC has played a role in the evolution and growth of the U.S. wind industry. Under existing law, wind projects placed in service on or after January 1, 2013, will not be eligible to receive the PTC incentive. Industry proponents are advocating for an extension of PTC availability, citing employment, economic development, and other considerations as justification for the extension. While a PTC extension may improve the prospects for U.S. wind development and manufacturing next year and beyond, the wind industry is influenced by a number of other factors. It is uncertain how the near- or long-term availability of the PTC incentive—in isolation of changes to other market factors—would either grow or sustain current wind development and manufacturing levels.

---

* This is an edited, reformatted and augmented version of Congressional Research Service, Publication No. R42576, dated June 20, 2012.

For 2012, the pending expiration of the wind PTC is actually creating a short-term surge in wind project development and related investment and employment. Wind installations in 2012 are expected to range somewhere between 10 to 12 gigawatts (GW)—a record year for the industry. However, market estimates for new installations in 2013 range from 1-2 GW if the PTC expires and 2-4 GW if the PTC is extended. Limited market activity in 2013 is partially explained by the uncertain nature of the PTC, which results in reduced manufacturing orders and development activity as developers and investors wait for official policy direction. Wind installation projections for 2014 and beyond vary with the assumed availability, and duration, of PTC incentives. However, all projections reviewed for this report expect annual U.S. wind turbine demand to be less than the existing U.S. turbine manufacturing capacity—approximately 13 GW per year.

Other factors that can affect wind development include (1) state renewable portfolio standards (RPS), (2) U.S. electricity demand growth, and (3) the price of natural gas. State RPS policies have been the primary demand creator for wind projects, in most cases, by requiring certain utilities to source a percentage of their retail electricity sales from renewable generators. Market analysis indicates that incremental RPS-driven demand for all sources of renewable power is estimated to be 4 GW-5 GW annually until 2025. Additionally, U.S. electricity demand growth is expected to be modest for the foreseeable future, meaning that there will likely be modest demand for new electric power capacity. Finally, the price of natural gas can also influence wind markets. Low natural gas prices can erode the economic competitiveness of wind electricity, while high natural gas prices can result in opportunities for wind to compete economically without the PTC. Current estimates from the U.S. Energy Information Administration (EIA) project sustained low, but increasing, natural gas prices for the next several years.

By the end of 2012, Congress will either allow the PTC incentive to expire or it may choose to extend or modify the incentive. Should Congress decide to extend the availability of wind PTC incentives, the duration (e.g., two years, four years, permanent) of such an extension will likely be part of the policy debate. Generally, the shorter the extension the greater the short-term economic and employment activity as developers and investors accelerate development plans in order to qualify for the PTC incentive. However, this development acceleration is likely to reduce future RPS-driven demand. A permanent PTC is also a policy option that may be considered, and EIA estimates indicate that such a policy may actually reduce near-term wind capacity additions, with annual installations peaking at 4 GW in the 2030 timeframe. Higher natural gas prices, more aggressive RPS policies, and increased U.S. electricity demand could change this outlook.

BACKGROUND

Federal incentives for operational renewable electricity projects have generally been in the form of tax benefits, including production tax credits (PTC), investment tax credits (ITC), and accelerated depreciation.[1] Renewable energy production tax credits were first introduced in the Energy Policy Act of 1992.[2] Section 45 of the Internal Revenue Code (IRC) outlines production tax credit incentives for wind, biomass, geothermal, landfill gas, trash, qualified hydropower, and marine and hydrokinetic projects that generate electricity. Under current law, the production tax credit for new wind projects will no longer be available as of January 1, 2013.[3] For all other eligible renewable energy projects, the PTC is available to projects placed in service before January 1, 2014.[4]

PTC policies provide incentives for electricity projects by providing a tax credit for each kilowatt-hour of electricity produced by a qualified project during the first 10 years of operation. Currently, the tax credit for wind projects is 2.2 cents ($0.022) per kilowatt-hour.[5] The PTC incentive is annually adjusted for inflation.

To date, the wind industry has been the largest beneficiary of federal production tax credits. The industry has experienced substantial growth over the last several years, with annual capacity installations generally increasing since 2005 (see **Figure 1**). As of the end of March 2012, cumulative U.S. wind power capacity was 48,611 megawatts, equal to approximately 4% of total U.S. generation capacity.[6] In 2011 wind was the largest source of non-hydro renewable electricity generation, providing approximately 120 million megawatt-hours, roughly 3% of total U.S. generation.[7] In response to U.S. wind market growth, a number of manufacturing and assembly facilities were established to supply wind turbine components and systems. In 2012, the U.S. wind manufacturing sector is estimated to have the capacity to produce approximately 13 gigawatts (GW) of wind turbines annually.[8] Industry estimates indicate that approximately 470 facilities in the United States provide various products (e.g., towers, turbines, gear boxes) for the wind turbine manufacturing supply chain.[9] In 2011, the wind industry reported that these facilities supported approximately 30,000 jobs.[10]

There are many arguments both for and against tax incentives for renewable electricity generation. Proponents of extending the wind PTC point to the potential loss of manufacturing and construction jobs that will result if the tax incentive is allowed to expire, the environmental benefits of U.S. wind development, and the potential to re-establish the United States as a global leader in an emerging industry. Opponents of a wind PTC extension argue that all electricity generators should be subject to market-based competition, wind electricity generation has been incentivized for a long enough period of time, and wind projects should compete on their own economic and environmental merits without the support of federal financial incentives.

This report examines how the production tax credit and its impending expiration impact the wind industry, and how other factors influence market demand for wind power projects.

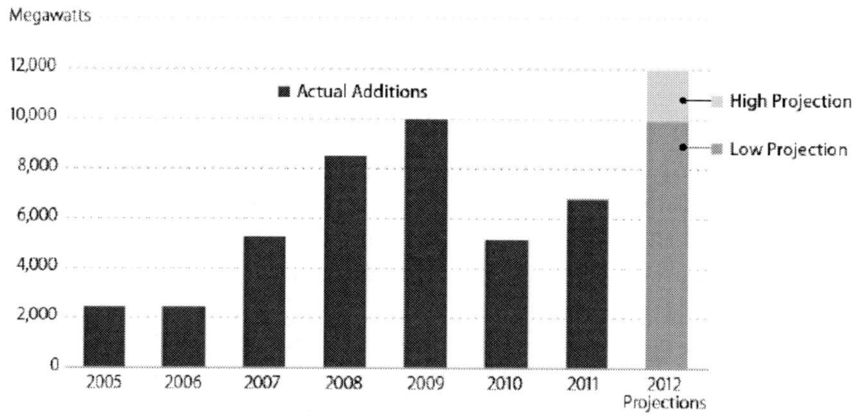

Source: American Wind Energy Association (AWEA) annual and quarterly reports. 2012 projections from Bloomberg New Energy Finance and CRS conversations with various wind market analysts.

Figure 1. Annual U.S. Wind Power Capacity Additions, (2005-2012).

IMPACT OF CURRENT PTC EXPIRATION

U.S. wind installations are expected to reach record levels in 2012, with industry analysts estimating between 10 GW and 12 GW of total installations by the end of the year.[11] As a result, related economic development, employment, and manufacturing activities needed to support 2012 wind installations are likely to be at record levels. Wind developers, utility companies, and investors are accelerating their planned wind projects in order to qualify for tax credit incentives that might not be available in 2013 and beyond. In essence, the pending PTC expiration at the end of 2012 has actually created a short-term surge in wind-related manufacturing and employment. Due to lead times—generally 12 to 18 months— required to fully develop wind projects, most manufacturing activities supporting 2012 wind capacity additions likely occurred either in 2011 or during the first quarter of 2012. Wind-related employment and economic development activity in the second half of 2012 will be primarily focused on construction, installation, and commissioning activities for projects in development. Based on current market conditions and other factors, it is unlikely that 2012 wind development levels can be sustained in either the near or long term, regardless of PTC availability.

Accelerated wind development in anticipation of the PTC expiring can create a severe market downturn in the year following PTC expiration. The wind PTC has expired three times since 2000 (in 2000, 2002, and 2004), and the wind industry experienced precipitous drops in annual wind capacity installations in each of those years.[12] One market estimate projects that 2013 wind capacity additions may drop to as low as 1 GW in 2013, if the PTC is not extended.[13]

POTENTIAL IMPACT OF EXTENDING THE PTC

Production tax credits for wind-generated electricity provide a financial incentive for project developers and investors to install wind projects in the United States. However, the PTC incentive is only one of several factors that influence wind development, and a PTC extension, in isolation of other market factors, may not result in ever-larger levels of wind deployment. Other important factors for project development include state renewable portfolio standards, electricity demand growth, and natural gas prices. Each of these factors is discussed in more detail below. The following sections provide some background on how a PTC extension might impact U.S. wind project installations and manufacturing. A brief discussion of the potential impact of a short-term versus long-term PTC extension is also provided.

Wind Installations

Various organizations have estimated the amount of wind project development under scenarios in which the PTC is either extended or is not extended. CRS obtained forecast information from the U.S. Energy Information Administration (EIA) and from Bloomberg New Energy Finance (BNEF) that estimates the amount of wind capacity (megawatts) expected to be installed under different PTC availability scenarios. **Figure 2** compares estimated wind installations with existing U.S. wind turbine manufacturing capacity.

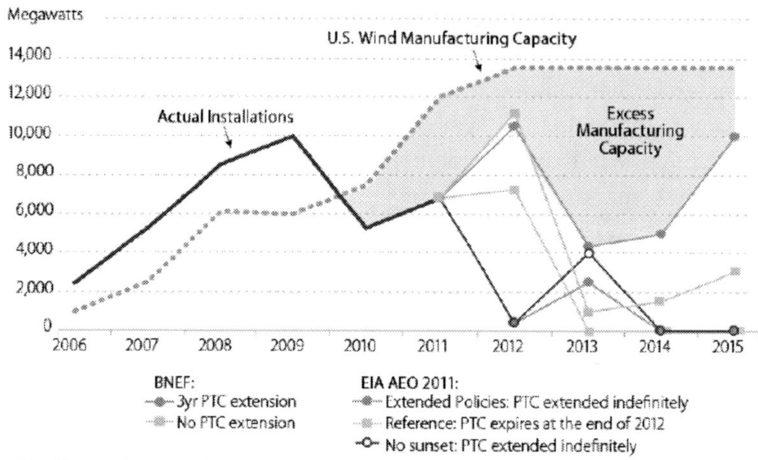

Source: Zindler, Ethan, "Clean Energy Market Trends: U.S. House Renewable Energy and Energy Efficiency Caucus," *Bloomberg New Energy Finance*, May 9, 2012; and, U.S. Energy Information Administration, *Annual Energy Outlook 2011*, AEO Table Browser, May 24, 2012, available at http://www.eia.gov/oiaf/aeo/ tablebrowser/ #release=AEO2011&subject=10-AEO2011&table=16-AEO2011& region=0-0&cases=nosunsetd030711a,extended-d031011a,ref2011-d020911a.

Notes: The BNEF "3yr PTC extension" case shows a large capacity addition increase from 2014 to 2015. This is likely explained by the expiring nature of the three-year extension and the expectation that developers would accelerate their projects in order to qualify for the PTC incentive before the deadline. Unlike the BNEF three-year extension scenario, EIA estimates wind capacity additions for two scenarios that assume indefinite availability of the PTC. As indicated in the figure, the long-term PTC extension actually results in less near-term development activity since there is no expiration date motivating developers to accelerate project development timelines. For background information about EIA's model and analysis results, see the "notes" section in Figure 3. EIA's "Extended Policies" and "No sunset" scenarios both assume an indefinite extension of PTC incentives, but annual capacity estimates are different for each scenario. In addition to the indefinite availability of PTC incentives, the "Extended Policies" scenario also includes assumptions for energy efficiency policies that are extended over longer periods of time. These energy efficiency measures effectively reduce U.S. electricity demand, and therefore reduce the amount of additional wind capacity needed to comply with RPS policies.

Figure 2. Estimated U.S. Wind Installations, With and Without a PTC Extension, (2006-2015).

As indicated in **Figure 2**, both BNEF and EIA forecast levels of wind development in 2013 and 2014 that are much less than development activity expected in 2012. As a result, levels of investment, manufacturing, and employment activity will be commensurately lower in the near term, even with a PTC extension.

However, a short-term PTC extension may result in higher near-term levels of development activity when compared to scenarios without a PTC extension.

Wind Turbine Manufacturing

Neither BNEF nor EIA estimate a scenario where wind installations meet or exceed existing U.S. wind turbine manufacturing capacity (see **Figure 2**). As a result, a PTC extension is unlikely to result in stimulating additional wind manufacturing facilities in the United States.

Estimated wind installations in 2013 and 2014 are expected to drop to levels much lower than existing U.S. manufacturing capacity, including PTC extension scenarios. Whether the PTC expires or is extended, U.S. wind manufacturing utilization levels will likely be less than levels needed to support the wind market in 2012. Therefore, some U.S. wind manufacturing facilities could reduce operations or even completely shut down in 2013 and beyond.

Much like the U.S. wind market, there is excess capacity in the global wind turbine manufacturing sector.[14] The competitive global market for wind generating equipment is one factor that may limit U.S. wind turbine manufacturing export opportunities.[15] Other factors affecting U.S. wind exports may include logistic and transportation costs associated with exporting large wind turbine equipment and certain local-content policies within global markets that may require co-locating manufacturing capability within a geographical market area.

However, excess wind turbine manufacturing capacity will likely result in wind turbine price decreases as manufacturers improve their cost and technology performance. Wind turbine price declines would contribute to new wind projects becoming more economically competitive with other sources of electricity generation on an unsubsidized basis.

Short-Term versus Long-Term PTC Extension

Some advocates for extending the availability of the PTC for wind projects argue that a long-term extension is needed to provide stable incentives that will result in certainty within the wind industry and may stimulate growth. The American Wind Energy Association (AWEA) states the following:

> The wind industry seeks long-term tax policies, lasting more than just a few years, to provide consistency and market certainty.[16]

AWEA and other proponents of extending the availability of the PTC incentive argue that the expiring nature of production tax credits has created a volatile U.S. wind market with new installations ramping up just before the credits expire, and the following year having very little new wind development.[17] It is possible that such uncertainty could reduce investment, research, and employment in the wind industry. Going forward, an element of the PTC debate may include the duration of PTC availability. If the PTC is extended, should it be a short-term, long-term, or indefinite extension?

Short-term PTC extensions generally result in short-term manufacturing, development, and employment activity as project developers and investors seek to capture the value of tax credit incentives during their availability window. However, since much of the demand for wind-generated electricity is a result of state-level renewable portfolio standards (discussed in more detail below), a short-term PTC extension would likely result in accelerating wind deployments needed to comply with state RPS requirements. This acceleration scenario is illustrated by the BNEF three-year PTC extension forecast (dark green line) in **Figure 2**, where annual installations reach 4 GW in 2013, 5 GW in 2014, and then ramp up to approximately 10 GW in 2015, when the credit extension would end. As a result, RPS-related demand in later years would likely decline and any future PTC extensions may or may not provide enough incentive to stimulate additional development activity.

Alternatively, a stable and long-term PTC incentive would provide manufacturers and developers with known incentive levels over an established period of time. However, a long-term or permanent PTC may not stimulate market activity comparable to levels observed between 2010 and 2012. EIA forecasts indicate that an indefinite PTC extension could result in more total wind capacity installations over the projection period (see **Figure 3**) when compared to a reference case scenario. However, annual capacity installations in the long-term extension scenario are relatively modest (in some years zero) and peak at around 4 GW. These annual installation levels are much lower than the existing 13 GW of U.S. wind turbine assembly capacity.

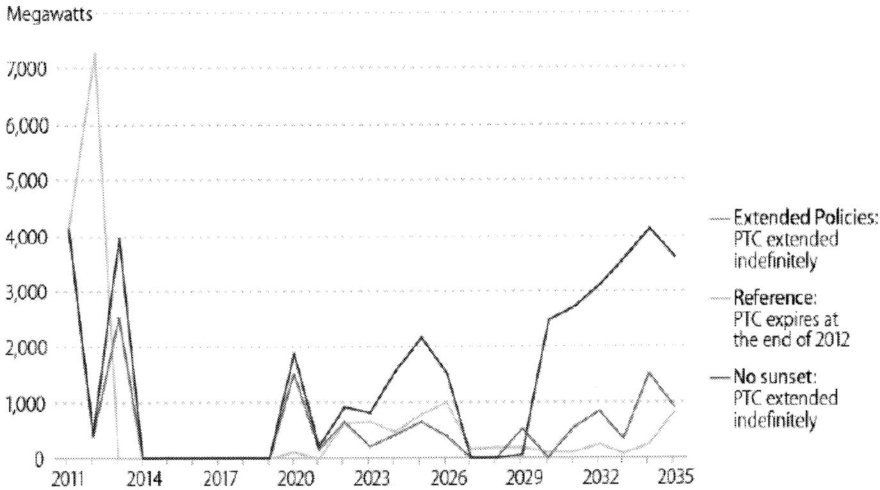

Source: U.S. Energy Information Administration, *Annual Energy Outlook 2011*, AEO Table Browser, May 24, 2012, available at http://www.eia.gov/oiaf/aeo/tablebrowser/#release=AEO2011&subject=10-AEO2011&table= 16-AEO2011&region=0-0&cases=nosunset-d030711a,extended-d031011a,ref2011-d020911a.

Notes: EIA uses its National Energy Modeling System (NEMS) to make annual projections for the U.S. energy sector. Annually installed wind capacity, as indicated in this figure, is one of several outputs from the NEMS model. NEMS includes several modules. One module is the Electricity Market module, which is used to calculate the projections provided in this figure. The Electricity Market module evaluates the U.S. electricity market based on different regions and considers future wind capacity additions based on regional economic and policy conditions. Generally, wind capacity is added based on federal policy, market economics, and as needed in order to comply with state RPS requirements. With respect to state RPS requirements, NEMS considers the amount of renewable electricity capacity in the various regions and the amount of renewable electricity required to comply with state RPS policies. NEMS does not, however, consider every policy design element included in each state RPS, such as credit sale limitations, credit banking, or credit borrowing.

Figure 3. EIA Estimates for Annual Wind Capacity Installations: Long-Term PTC, (2011-2035).

A Note About Market Forecasts and Analytical Models

Market analysts have a difficult job in that they are required to predict the future in order to forecast expected market activity. Typically, market forecasts are estimated using a predictive numerical model that includes a number of assumptions about various factors that can significantly impact the forecasted results. Generally speaking, near-term forecasts have a higher degree of accuracy than long-term forecasts, simply because assumptions and variables that influence the model are more difficult to predict over longer time frames. Furthermore, forecasts will change over time as market conditions and other variables change.

Forecasts and predictions referenced in this report are no exception. Two different forecasting methodologies, with different assumptions, were used to derive the forecasts in this report, and results from each approach are obviously different. As a result, each estimate varies and neither forecast can be considered correct or incorrect. Additionally, estimates provided in this report are static as of the date they were released. In practice, market forecasts are typically updated periodically to reflect market, financial, and policy changes.

Market forecasts are valuable to policy makers, project developers, manufacturers, and other stakeholders since they allow for an assessment of expected market activity under certain assumptions. Furthermore, market forecasts can provide some perspective on the potential impact of market and policy changes. Nevertheless, it is important to recognize that models used to forecast the future have limits and reality will most likely differ from the projections provided in this report. Finally, the level of accuracy of market forecasts can only be determined by observing actual results in the future.

OTHER FACTORS THAT AFFECT U.S. WIND DEVELOPMENT

As briefly discussed above, availability of the federal production tax credit is one of several factors that impact the amount of wind development and deployment in the United States. Generally, state-level renewable portfolio standards (RPS) create a source of demand for wind projects. Overall U.S. electricity demand growth is also an important factor as it determines the total addressable market that wind projects can target. Low natural gas prices can create economic competitiveness pressure for wind projects but high natural gas prices can result in additional opportunities for the wind sector. While not an all-inclusive list of factors that affect wind development, the factors addressed below do represent some of the critical non-PTC factors that influence the U.S. wind industry. The following sections provide additional details about each factor.

State Renewable Portfolio Standards

Generally, but not in all instances, a renewable portfolio standard is a policy that requires a certain percentage of electricity sold or generated within a defined geographical area be derived from qualified renewable energy sources.[18] As of May 2012, 29 states plus the District of Columbia and Puerto Rico had binding RPS policies.[19] While the general concept of an RPS is the same for all states, each state typically has a unique design and implementation approach for its respective RPS policy. For example, the state of California requires that by 2020 its utility companies will have 33% of their retail electricity sales generated from renewable energy sources.[20]

State RPS policies are the primary renewable electricity demand driver, although demand for renewable power can also be encouraged by voluntary green power programs and fundamental economics.[21] Analysis by Lawrence Berkeley National Laboratory (LBNL) indicates that approximately 27 GW of non-hydro renewable electricity capacity was added in states with RPS policies in the years 1998-2010.[22] On a capacity basis, 92% of these renewable electricity additions were wind power projects.[23] From a generation perspective, the combination of mandated demand at the state level and federal financial incentives has created an environment that supports development of renewable electricity projects, most notably wind projects.

One typical compliance approach for RPS policies is submitting renewable energy certificates (RECs) to the appropriate state agency that manages RPS compliance.[24] RECs, each of which receives a unique tracking identification number, represent the renewable attributes of electricity generated from a qualified renewable power facility. One REC typically represents one megawatt-hour of renewable electricity. RECs can be obtained on either a bundled basis, where a utility company contracts to purchase both the electricity and associated RECs from a renewable generator, or an unbundled basis, in which case a utility company may purchase qualified

RECs from other entities. RECs can potentially provide an additional revenue source for wind projects, although the value of RECs can vary depending on the supply/demand balance within certain markets.

Analysis of state RPS compliance indicates that existing renewable electricity capacity may be adequate to allow for RPS compliance over the next several years. Furthermore, future RPSdriven demand may not be large enough to stimulate substantial growth in the wind electricity sector. LBNL estimates that approximately 4 GW-5 GW of annual renewable electricity capacity additions between 2011 and 2025 would be required in order to meet state RPS requirements.[25]

This compares to 6 GW-11 GW of renewable capacity installed between 2008 and 2010.[26] RPSdriven renewable electricity demand, on its own, does not appear large enough to support annual wind industry growth going forward, and RPS policies may not provide enough demand to sustain annual wind capacity installations compared to levels installed in the years 2009 to 2012, or to match current U.S. manufacturing capacity.

U.S. Electricity Demand Growth

Electricity demand growth is an important factor when considering opportunities for renewable electricity for two primary reasons. First, generally, the greater the annual demand growth the more new electricity capacity needed to satisfy that demand. Larger annual requirements for new electricity capacity create more opportunities for renewable electricity projects to compete. Second, large annual demand growth can result in a larger base of electricity to which RPS policies are applied. The larger the electricity base, the greater the amount of renewable electricity required to comply with state percentage-based RPS policies. However, EIA projects modest growth levels for U.S. electricity demand over the next several years (see **Figure 4**).

Long-term U.S. electricity demand is expected to continue along a modest annual growth path out to 2035. However, electric power demand in developing economies, such as China, is expected to experience significant annual growth out to 2035 (see **Figure 5**).

As indicated in **Figure 5**, expected electric power demand growth profiles for the United States and China are very different, with China's growth level forecasted to be much larger than that of the United States. Therefore, opportunities for new electric generation capacity in China will be commensurately larger. As result, the Chinese electricity market may present more opportunities for renewable electricity projects, including wind power, due to the large amount of additional installations needed to meet projected electricity demand.

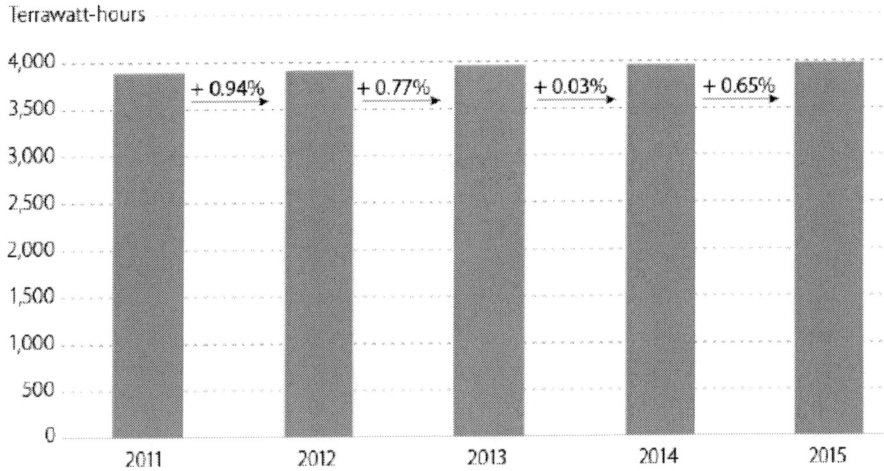

Source: *EIA Annual Energy Outlook 2011*, U.S. Energy Information Administration.

Figure 4. Projected U.S. Electricity Demand Growth, (2011-2015).

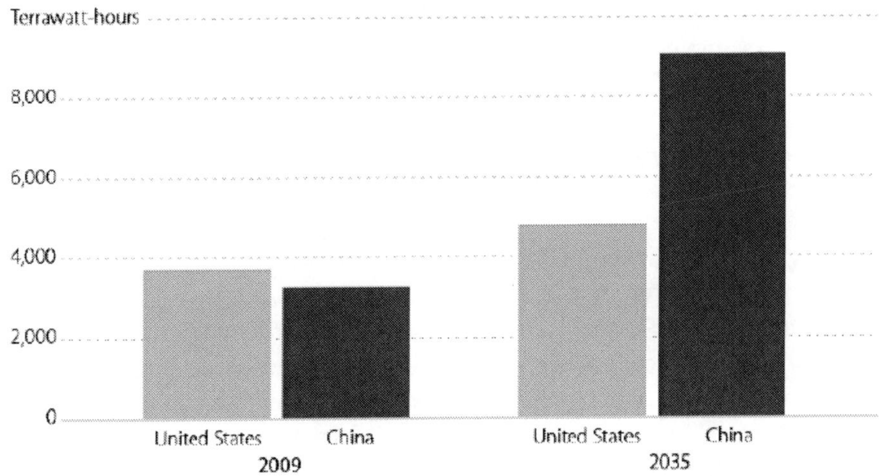

Source: *World Energy Outlook 2011*, International Energy Agency, 2011.

Figure 5. U.S. and Chinese Projected Electricity Demand Growth, (Annual Demand in 2009 and 2035).

Natural Gas Price

The price of natural gas also has an impact on the U.S. wind market. Generally, lower natural gas prices can reduce the economic competitiveness of wind power, while higher natural gas prices can create opportunities for wind to compete on economics alone, in some cases without subsidies. Since wind power economics vary depending on project location, there is no single natural gas price level at which all wind projects can compete either on an unsubsidized basis or with the availability of PTC incentives. Furthermore, natural gas prices can affect wind power in different ways depending on the state or region in which a wind project operates.

U.S. electricity markets are complex, and a comprehensive analysis of electricity markets is beyond the scope of this report.[27] Generally, however, there are two distinct types of markets in the United States: (1) competitive markets: power generators are subject to price competition when selling power into wholesale markets, and (2) cost-of-service markets: power generators earn a regulated rate of return established by a public utilities commission.[28] According to one estimate, approximately two-thirds of electricity consumed in the United States is within competitive markets.[29] Furthermore, there are several regional power markets in the United States, each with a unique market structure, fuel mix, and set of rules that govern market operations. Depending on the respective market characteristics, natural gas prices can impact wind projects in different ways. The following sections provide two simplified examples of how natural gas prices might impact the economics, and development, of U.S. wind power projects.

Example 1: Markets Coordinated by a Regional Transmission Organization

Competitive electricity markets are typically managed by a Regional Transmission Organization (RTO) or an Independent System Operator (ISO), a third-party operator of the electricity transmission system for a defined geographical area. In essence, the RTO provides a market making function and is a critical interface between electricity purchasers and suppliers. RTO-coordinated markets can generally be described as markets where wholesale electricity rates are frequently established (typically on an hourly basis) through a bidding process. Power generators provide bids, usually based on the variable cost for each respective generator, to supply a certain amount of electricity. The RTO will organize the bids from the lowest to the highest. The bid offer price that matches the level of electricity supply necessary to meet power demand sets what is known as the "clearing price." **Figure 6** provides a simplified example of how the clearing price might be established for wholesale electric power within an RTO-coordinated market. All generators that supply electricity at or below the clearing price are paid for their electricity supply at the clearing price level. However, many power generators may establish power purchase agreements (PPAs) directly with utility companies to provide long-term revenue certainty. In these instances financial transactions between generators and power purchasers will often occur exclusive of the RTO clearing price mechanism in order to satisfy PPA terms and conditions.

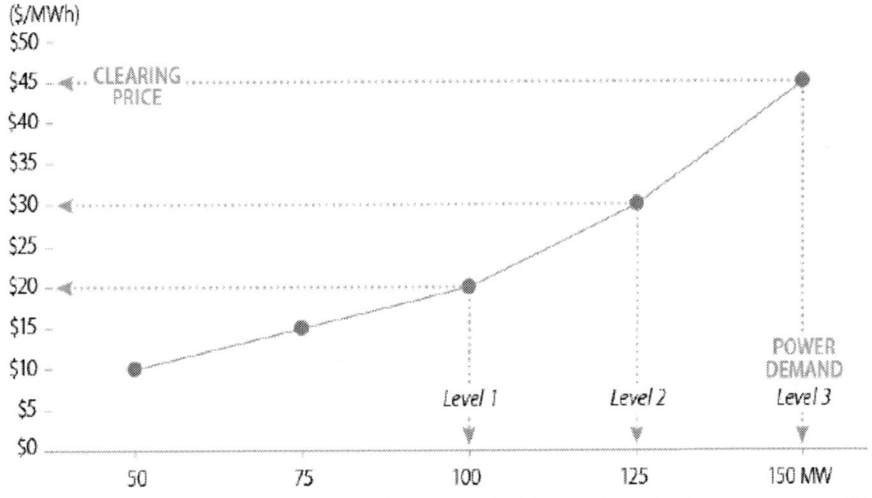

Source: CRS adaptation of *Locational Marginal Pricing (LMP) Overview*, page 22, PJM, October 18, 2011, available at https://www.pjm.com/training/~/media training/core-curriculum/ip-lse-201/lmp-overview.ashx.

Notes: This figure illustrates how the wholesale electricity clearing price might change as a function of power demand. The red line represents the bid offer prices for electricity that are organized from low to high. Depending on the level of demand (three different hypothetical levels illustrated in this figure), the clearing price is adjusted in order to satisfy required demand for electricity delivery during a certain time period.

MW = megawatts

MWh = megawatt-hour.

Figure 6. Simplified Electricity Dispatch Curve for Wholesale Power, (Hypothetical RTO Market Example).

In certain electricity markets, during different times of year, and during certain times throughout a day, especially during daytime hours when electricity demand peaks, natural gas power generation sets the clearing price. Since natural gas fuel costs are the largest contributor to natural gas power generation costs, there will be some degree of correlation between the price of natural gas and the wholesale electricity clearing price within certain markets. Generally, as natural gas prices rise, so does the clearing price during certain times throughout the day. However, total electricity demand within a market can also impact wholesale electricity prices. For wind projects that participate in this type of market without a PPA, also known as "merchant wind," the clearing price will usually determine the revenue received for electricity sold into the market. Although, in certain instances, wind projects can supplement their electric power revenue by selling renewable energy certificates (RECs) to entities required to comply with state RPS policies.[30] Nevertheless, higher natural gas prices and the resulting higher electricity clearing prices can increase revenues for wind projects thereby making them attractive investment and development opportunities. Conversely, lower natural gas prices and lower clearing prices can decrease wind project revenues to a point where projects are not economically viable.

Wind projects in RTO-coordinated markets can mitigate wholesale market price risk by entering into long term PPAs with utility companies. In this case, utility companies absorb the risks associated with low wholesale clearing prices. Utilities may be motivated to enter into PPAs with wind projects as a means of complying with state RPS policies or as a way to hedge against rising natural gas and wholesale electricity prices. However, many state RPS policies include an alternative compliance payment (ACP) design element whereby utilities can opt to make payments to an ACP fund instead of generating or purchasing a required amount of renewable electricity. Low natural gas prices can lower electricity prices and result in making ACPs more economical than either building or paying for renewable generation. Thus, the short- and longterm price of natural gas, along with any ACP policy, can impact a utility company's decision to enter into PPAs with renewable electricity generators.

Example 2: California RPS Cost Containment Approach

The state of California currently has one of the most aggressive RPS policies.[31] However, as part of the policy design, the California RPS includes a cost containment design element, which is directly linked to the price of natural gas. California has used a Market Price Referent (MPR) as a benchmark for determining the price premium required to support certain sources of renewable electricity. If contract prices for renewable electricity exceed MPR levels, then formal approval by the California Public Utilities Commission (CPUC) of the contract must be obtained and above market funds (AMFs) must be available to compensate for the additional cost associated with purchasing the renewable power.[32] AMFs establish cost limits for California electric utility companies required to comply with the state's RPS policy.[33] Benchmark MPR prices are set based on the levelized price of electricity from a 500 MW natural gas-fired combined cycle gas turbine (CCGT).[34] Consequently, natural gas prices can significantly influence MPR benchmark price levels. This approach is designed to contain costs associated with RPS implementation since "the MPR sets a limit on the procurement obligations of retail sellers under the RPS program."[35]

POLICY DISCUSSION

The 112th Congress may decide if the PTC incentive for wind electricity will be extended, modified, or terminated. During the congressional debate about the future of the wind PTC incentive, Congress may consider various policy options, including those discussed below.

Allow the PTC to Expire

Absent congressional action, the PTC incentive for wind electricity projects will no longer be available for new installations placed in service after January 1, 2013. Some market projections suggest that annual wind capacity additions will decline precipitously if the PTC expires (see **Figure 2**). As a result, wind-related manufacturing and project development employment would decline as well. Allowing the PTC to expire may motivate wind equipment manufacturers and developers to take certain actions (e.g., maximize turbine performance, minimize manufacturing costs) necessary to make wind electricity more broadly competitive on an unsubsidized basis. These actions could potentially result in a stronger and more robust, although possibly smaller, wind industry that can compete directly with all sources of power generation. However if state RPS policies remain as-is and low natural gas prices persist, a prolonged industry contraction could limit the ability of the wind industry to respond once, and if, market conditions change.

Extend the PTC Incentive

Congress might also consider extending the PTC incentive. Some market estimates indicate that a PTC extension would result in increasing U.S. wind capacity installations, when compared with allowing the PTC to expire—but at levels less than those observed since 2009 and less than current U.S. wind turbine manufacturing capacity (see Figure 2). Determining the duration of a possible PTC extension is also an important policy consideration. Generally, the shorter the extension the more near-term wind market activity that may be stimulated since project developers are motivated to install new capacity in order to qualify for PTC incentives.

Table 1. Selected Bills Introduced to Extend Wind PTC Incentives (112th Congress)

Bill Number	Bill Title	PTC Extension	Status
H.R. 3307	American Renewable Energy Production Tax Credit Extension Act of 2011	4 years	November 2, 2011; referred to House Committee on Ways and Means
H.R. 5187	IMPACT Act of 2012	8 years	April 27, 2012; referred to House Energy and Commerce Subcommittee on Energy and Power
S.Amdt. 1812	Amendment to S. 1813 (MAP-21)	1 year	March 13, 2012; not agreed to by a vote of 49-49
S. 2201	American Energy and Job Promotion Act	2 years	March 15, 2012; referred to Senate Finance Committee
S. 2204	Repeal Big Oil Tax Subsidies Act	1 year	March 29, 2012; Senate voted 51-47 to not invoke cloture

Source: Legislative Information System.

However, depending on the timing of an extension, a one-year extension may have limited impact due to 12-18 month wind project development lead times. Also, as discussed above, the near-term market stimulation that might result from a PTC extension may accelerate wind development at the expense of future-year RPS-driven demand. A permanent PTC may not stimulate near-term wind development activity since there would likely be less motivation to accelerate projects in order qualify for federal tax incentives (see Figure 3). However, in a market where natural gas prices are rising, a permanent PTC could potentially stimulate wind electricity demand that might not otherwise occur. Some bills have been introduced in the 112th Congress that would extend the availability of the PTC incentive for wind electricity projects. Table 1 includes four bills that have been introduced and compares the duration of the PTC extension under each proposal.

Phase-Out of the PTC

Another policy option that has been discussed during the 112th Congress is the possibility of an extension that phases out the PTC incentive over time.[36] The basic concept of a PTC phase-out is to gradually reduce the value of the incentive over time in order to provide the industry a degree of certainty and a motivation to improve cost and performance efficiencies in order to become price competitive without the PTC.[37] The design of a PTC phase-out policy could potentially be difficult because, in order to stimulate U.S. wind development, the rate at which the PTC is reduced may need to be offset by, or aligned to changes in, other market factors (e.g., higher natural gas prices, more stringent state RPS policies, increased U.S. electricity demand). These other market factors will likely be beyond the scope of control of the PTC phase-out policy. An alternative approach may be a dynamic PTC phase-out design that requires a minimum PTC incentive reduction annually but could also be adjusted based on other market conditions (e.g., natural gas prices, system costs, technology improvements). Implementation of a dynamic PTC phase-out could potentially be complicated as well and the policy would need to be designed in such a way to motivate the industry to continue cost reduction and technology improvement initiatives.

End Notes

[1] For additional background on U.S. energy tax policy and the production tax credit for renewable energy projects, see CRS Report R41227, *Energy Tax Policy: Historical Perspectives on and Current Status of Energy Tax Expenditures*, by Molly F. Sherlock.

[2] Ibid.

[3] Production tax credits for wind power projects are for those that use wind turbines larger than 100 kilowatts (the majority of current installed capacity). For wind projects that use 100 kilowatt and smaller wind turbines, a 30% investment tax credit is available until January 1, 2017.

[4] For more information see IRC §45.

[5] "Credit for Renewable Electricity Production, Refined Coal Production, and Indian Coal Production, and Publication of Inflation Adjustment Factors and Reference Prices for

Calendar Year 2012," Department of the Treasury: Internal Revenue Service, Federal Register Vol. 77, No. 70, p. 21835, April 11, 2012.

[6] *AWEA U.S. Wind Industry First Quarter 2012 Market Report*, American Wind Energy Association, 2012, available at http://www.awea.org /learnabout/publications/reports/ upload/AWEA_First_Quarter_2012 _Market_Report_Public.pdf.

[7] *Electric Power Monthly*, U.S. Energy Information Administration, March 2012.

[8] Zindler, Ethan. "Clean Energy Market Trends: US House Renewable Energy and Energy Efficiency Caucus," *Bloomberg New Energy Finance*, May 9, 2012.

[9] AWEA U.S. Wind Industry Annual Market Report: Year Ending 2011, American Wind Energy Association, 2012. Also, for more information about U.S. wind manufacturing, see CRS Report R42023, U.S. Wind Turbine Manufacturing: Federal Support for an Emerging Industry, by Michaela D. Platzer.

[10] Ibid.

[11] Bloomberg New Energy Finance estimates 2012 wind installations to be approximately 10.5 gigawatts; see Zindler, Ethan, "Clean Energy Market Trends: US House Renewable Energy and Energy Efficiency Caucus," *Bloomberg New Energy Finance*, May 9, 2012. CRS conversations with other market analysts indicate that 10 GW to 12 GW in 2012 is a general range for wind capacity installations.

[12] Federal Production Tax Credit for Wind Energy: The American Wind Industry Urges Congress to Take Immediate Action to Pass an Extension of the PTC, American Wind Energy Association, available at http://www.awea.org/issues/ federal_policy/upload/PTC-Fact-Sheet.pdf.

[13] Zindler, op. cit.

[14] "Q1 2012 Clean Energy Policy & Market Briefing: Policy pullbacks, weak economies, market oversupply slow Q1 investment," *Bloomberg New Energy Finance*, April 18, 2012.

[15] According to the Global Trade Information Service, U.S. wind power generating set (HS 8502.31) exports were approximately $255 million in 2011.

[16] American Wind Energy Association website, June 4, 2012, http://www.awea.org/issues/ federal_policy/index.cfm.

[17] Federal Production Tax Credit for Wind Energy: The American Wind Industry Urges Congress to Take Immediate Action to Pass an Extension of the PTC, American Wind Energy Association, available at http://www.awea.org/issues/ federal_policy/upload/PTC-Fact-Sheet.pdf.

[18] The state of Texas is an exception to this generality. According to DSIRE (Database of State Incentives for Renewables & Efficiency, http://www.dsireusa.org), Texas has a renewable generation requirement that requires 5,880MW of installed renewable capacity by 2015.

[19] Database of State Incentives for Renewables & Efficiency, http://www.dsireusa.org. A summary map of state RPS policies is available at http://www.dsireusa.org/documents/ summarymaps /RPS_map.pdf.

[20] Database of State Incentives for Renewables & Efficiency, available at http://www.dsireusa.org/incentives/ incentive.cfm?Incentive_Code=CA25R&re=1&ee=1.

[21] Voluntary green power markets are those in which consumers, businesses, and other entities purchase a certain amount of renewable energy on a voluntary basis. For more information on voluntary green power, see Jenny Heeter and Lori Bird, "Status and Trends in U.S. Compliance and Voluntary Renewable Energy Certificate Markets (2010 Data)," National Renewable Energy Laboratory, October 2011.

[22] Ryan Wiser and Galen Barbose, "The State of the States: Update on the Implementation of U.S. Renewable Portfolio Standards," Presentation at the 2011 National Summit on RPS, Lawrence Berkeley National Laboratory, October 26, 2011.

[23] Ibid.

[24] For additional background on RECs, see Holt, E., Sumner, J., and Bird, L., *Role of Renewable Energy Certificates in Developing New Renewable Energy Projects*, National Renewable Energy Laboratory, 2011, available at http://www.nrel.gov/docs/fy11osti/51904.pdf.

[25] Ryan Wiser and Galen Barbose, "The State of the States: Update on the Implementation of U.S. Renewable Portfolio Standards," Presentation at the 2011 National Summit on RPS, Lawrence Berkeley National Laboratory, October 26, 2011.

[26] Ibid.

[27] For additional background on the U.S. power sector, see *Electricity Primer–The Basics of Power and Competitive Markets*, Electricity Power Supply Association, available at http://www.epsa.org/industry/primer/.

[28] Ibid.

[29] ISO/RTO Council, http://www.isorto.org/site/c.jhKQIZPBImE/b.2603917 /k.B00F/About.htm, June 20, 2012.

[30] Renewable Energy Credits (RECs) are issued to renewable electricity generators as a means of documenting, and accounting for, the renewable energy attributes of power generation. In some markets, RECs are "bundled" with electric power, in which case the purchaser of the power also receives the RECs. In other markets RECS are "unbundled," in which case the electric power and the RECs may be sold to different buyers. For additional background on RECs, see Holt, E., Sumner, J., and Bird, L., *Role of Renewable Energy Certificates in Developing New Renewable Energy Projects*, National Renewable Energy Laboratory, 2011, available at http://www.nrel.gov/docs/ fy11osti/51904.pdf.

[31] For additional information about California's RPS policy, see http://www.dsireusa.org/ incentives/incentive.cfm? Incentive_Code =CA25R&re=1&ee=1.

[32] Additional information about California's Market Price Referent is available at http://www.cpuc.ca.gov/PUC/energy/ Renewables/mpr.

[33] For more information about California's approach to RPS Cost Containment, see http://www.cpuc.ca.gov/PUC/ energy/Renewables /SB1036implementation.htm, June 11, 2012.

[34] The term "levelized price" basically reflects the average price at which the baseline CCGT power generation plant would need to sell electricity in order to pay for capital, operation, maintenance, fuel, and finance costs over a defined period of time. A critical assumption used for calculating levelized price is the power plant capacity factor, the amount of operating time during each calendar year.

[35] *Resolution E-4442*, Public Utilities Commission of the State of California, December 1, 2011, available at http://docs.cpuc.ca.gov/WORD_PDF /FINAL_RESOLUTION/154753.PDF.

[36] Nick Juliano, "Interest grows in Senate for phaseout of production tax credit," *E&E Daily*, March 28, 2012.

[37] Under current law (IRC Section 45), the PTC includes a phase-out provision that is based on a reference electricity price. Details of the phase-out existing phase-out provision are included in IRC Section 45, available at http://www.novoco.com/energy/resource_files/irs_guidance/ irc/section_45.pdf.

INDEX

D

E

F

M

N

O

P

R

S

T

U

V

W